Regionalization of the World

SCIENCES

Geography and Demography, Field Director – Denise Pumain

The World in its Divisions: Borders and Discontinuities,
Subject Head – Clarisse Didelon-Loiseau

Regionalization of the World

Comparing Regional Integrations

Coordinated by
Pierre Beckouche
Yann Richard

iSTE

WILEY

First published 2024 in Great Britain and the United States by ISTE Ltd and John Wiley & Sons, Inc.

ISTE Ltd
27-37 St George's Road
London SW19 4EU
UK

www.iste.co.uk

John Wiley & Sons, Inc.
111 River Street
Hoboken, NJ 07030
USA

www.wiley.com

Library of Congress Control Number: 2023943507

British Library Cataloguing-in-Publication Data
A CIP record for this book is available from the British Library
ISBN 978-1-78945-158-0

ERC code:
SH2 Institutions, Values, Environment and Space
SH2_11 Human, economic and social geography

Contents

Introduction

Pierre BECKOUCHE[1] **and Yann RICHARD**[2]
[1] *UMR Ladyss, Université Paris 1 Panthéon-Sorbonne, France*
[2] *UMR Prodig, Université Paris 1 Panthéon-Sorbonne, France*

Following the global financial crisis that arose from the subprime crisis in the United States (2008), after "America first" witnessed during Trump's presidency in the United States (2017–2020) and after the Covid-19 pandemic (2020–2022), the globalization paradigm lost its omnipotence. At the economic level, the theme of "deglobalization" advocates the relocation of some activities (Latouche 2019), a priority for local production intended for the national market – especially in the countries of the South – in place of imports (Bello 2002), even protectionism (Sapir 2010), and the regulation of financial activities. At the political level, this takes the form of sovereignty and sometimes a revival of nationalism. When globalization is challenged, it is often in the name of local distribution channels that allow local actors to control their future, preserve their jobs and their environment, and/or in the name of a return to national sovereignty.

We might have thought that these demands for more controlled internationalization would lead to the promotion of a world organized in large international regions, more capable of regulating things than on the scale of the vast world. Nevertheless, the first two decades of the 21st century have not seen the victory of the regions. Instead, we hear more about the local and the return of nationalism. As for globalization, it is far from having surrendered: space is now structured by a planetary connection; the world's metropolises form a circumterrestrial archipelago animated by flows that never stop; the power of multinationals remains huge.

There are even signs of regional disintegration here and there. This was the case in Europe, at least until the Russian–Ukrainian war that broke out just as this book was being written. All Europeans have suddenly become geographers, asking themselves whether political Europe should stretch as far as Ukraine, whether the Donbass should be part of Western Europe rather than a Russian regional entity, and whether the European Union should integrate militarily. But it should be remembered that before then, the EU project had suffered a dramatic setback with the withdrawal of the United Kingdom (Brexit vote in 2016). Some EU member countries were turning inward and rejecting the community logic. We could speak of populist, Eurosceptic or sovereignist governments, but this also reflected a dislike of the European Union among citizens, the first signs of which were seen in the 1990s. These years also saw the end of the Soviet Union, the Warsaw Pact and the Council for Mutual Economic Assistance, the other great European regional experiment.

Regional uncertainties are not unique to Europe. In the Americas, MERCOSUR and UNASUR are stalled by recurrent economic and political crises, a structural lack of trust among some member countries and inadequate infrastructure. Venezuela's exclusion from MERCOSUR is a sign of a Latin American regionalism that seems to accumulate institutions without coherence. In East Asia, the disputes between China, Taiwan, Japan, North Korea and other countries in the region remain immense. In Africa, large regions are having difficulty driving development. We wonder whether the very idea of a large region is not simply being swept away by a Sino-American bipolarization of world space, shaking up the logic of proximity through international mega-treaties and Belt and Road initiatives.

However, the long-term trends are obstinate. States are giving more and more importance to relations with their neighbors. This is reflected in the spectacular increase in the number of regional agreements ("regionalism") for trade in goods and services. Today, 450 such agreements, notified to the WTO, are active. In 2021, the free trade area of the 44 members of the African Union came into force (African Continental Free Trade Area, AfCFTA). In January 2022, the world's largest free trade area, the Regional Comprehensive Economic Partnership (RECP), involving East and Southeast Asian countries, came into effect as China's response to the US attempts to create a trans-Pacific treaty. For the past 20 years, Russia has been trying to strengthen ties with its neighbors – former Soviet republics – through the Eurasian Economic Union and the Collective Security Treaty Organization, the influence of the Runet and, more recently, through arms. The Covid-19 pandemic and the rise of the importance of the environment and increased climate change to the top of the political agenda are prompting many actors to seek solutions at the scale of their regional neighborhood.

This regionalism is based on a major trend that globalization had masked: the regionalization of trade. For the past 50 years, many interactions have been concentrated in clusters of neighboring countries, whether in trade, investment, international mobility or the joint management of transboundary environmental public goods. These trends have long been observed by specialists from different disciplines who are interested in the effects of geographical proximity or in the comparative advantages and disadvantages of multi-state regionalism. Looking at the existing bibliography, economists and political scientists have dominated the debate for several decades. Geographers are also interested in the subject, looking at objects as diverse as the relations between metropolises, North/South regionalism, European, Latin American or East Asian integration, mobilities or international migrations.

Several recent events have revived thinking about the European region. Firstly, Brexit is proving more difficult than was envisaged by British Eurosceptics. Undoing ties that have been built over decades is a daunting exercise. Along the way, we realize that there is much to lose and that signing a separation agreement does not instantly sweep away the functional integration that has been built on the ground. Secondly, several countries are still knocking on the door of the European Union (recently Georgia, Moldova and Ukraine). Thirdly, Brexit does not seem to set a precedent, as there are no further requests to leave the European Union. Fourthly, European institutions and member countries continue to build the EU on a day-to-day basis. Research shows that regionalism is still very much in vogue, because it echoes major economic (construction of large markets, regulation of trade), political (international regulation of globalization, co-construction of norms and standards) and geopolitical (pacification of relations between neighboring states and structuring of a multipolar world, not to mention the work that will be devoted to the new European geopolitics which arose on February 24, 2022) challenges.

On the world map, despite the headwinds, the major regional groupings are there and there they are indeed. They have gradually imposed themselves, with varying degrees of success and following different trajectories. The first objective of this book is to answer the following question: have the regions become an additional level in the spatial deployment of numerous actors (political actors, firms, social groups, cultural organizations)? They could thus come to be inserted between the national and the global levels: the continuous enlargement of the scales of exchange, environmental problems, migratory routes, energy distribution, construction of major infrastructures. All of this goes beyond national limits and leads to regional cooperation. A second hypothesis: as more than just levels, these large entities become real territories, appropriated by the societies that constitute them, subject to shared representations, steered by powers that are increasingly defined and even increasingly competent on this macro-regional scale. In this hypothesis, which corresponds to what economists call "deep" regional integration, growing integration at the economic, institutional, legal and normative, political, cultural and identity levels make the region a landmark that makes sense, takes root in representations and models what could become a regional society.

The second objective of this book is to show that geographers are equipped to test these hypotheses and to better understand these large regions, even though they have long remained discreet in this field of research dominated by economics, political science and international law. Their legitimacy in the study of regions, scales, territories and spatial integration is not in doubt. They are aware of the contributions of other disciplines in order to shed light on their own research, even though these studies still too often lack the necessary generalization to go beyond the stage of regional monographs.

The first chapter of this book lays francophone geography's historical and theoretical contribution to the large regional issue.

Chapters 2–8 are thematic approaches to regional geography. The aim is to observe and even measure the regional concentration of certain practices and to see whether it combines with or opposes globalization. It is also a question of seeing whether this scale is not the most relevant for dealing with certain issues, in fields such as security, energy, planning, transport and even finance, since cooperation on a regional scale is easier when neighboring societies share certain collective preferences and are linked by strong interdependencies.

– Chapter 2 studies international migration and mobility. Camille Schmoll shows that migration routes, although increasingly global, retain a predominantly regional dimension.

– In Chapter 3, Angélique Palle looks at regional energy integration, which she observes in different parts of the world. She explains that the promoters of such integration – with its technical, political and economic components – are wondering what would be the optimal geographical scale for the construction of energy systems, particularly in the perspective of the transition to less carbon-intensive economies. Are large multi-state entities the optimal scale?

– In Chapter 4, Antoine Beyer describes the strong mutual links between regional integration and the dynamics of transport networks, by observing the flows, infrastructures and institutions that frame them. According to him, transport networks are markers of the interdependencies built up between neighboring countries.

– In Chapter 5, Gilles Van Hamme takes stock of the global dynamics of international trade in order to examine the regional fact. He hypothesizes that globalization and regionalization are not incompatible. Rather, they are concomitant and interrelated.

– In Chapter 6, Maude Sainteville analyzes financial exchanges, which are essentially dematerialized and globalized. In a counter-intuitive way, she shows that there is a regional dimension to the organization and practices of stock market activity, by focusing on three themes: the listing of securities, the construction of stock market groups and high-frequency trading.

– Kevin Limonier opens Chapter 7 with the obvious: digital flows are by nature ubiquitous and globalized. But by adopting a geopolitical perspective through the example of the *Runet* – i.e. all the networks, digital services and platforms where the Russian language is used to communicate – he indicates that these flows do not escape regionalization either.

– Chapter 8, written by Emmanuel Chauvin, focuses on cross-border security issues, an essential component of today's regions. He observes both the regional construction of security and the regional concentration of insecurity in regional conflict complexes. In his view, the "regional" scale must be considered with finesse: regional security architectures are often macro-regional in size, while the geography of violence is often concentrated in smaller aggregates of neighboring countries.

Chapters 9–16 approach regional integration in different parts of the world: Europe, Africa, East and Southeast Asia, North and South America, and the Arctic. The authors answer two questions in each chapter. What are the regional stakes in various large groups, can we speak of increasing intraregional flows ("regionalization") in all parts of the world? Is the awareness of these facts sufficiently shared to lead to real policies of regional integration ("regionalism")?

– Chapter 9 provides a picture of the uneven success of regional integration in Africa. Géraud Magrin and Olivier Ninot analyze the tensions between recent integrations driven by official institutions, and older functional cross-border integrations "from below". Given the fragility of the states concerned, they ask to what extent regional integration can be detrimental to state-building and hence to development.

– By adopting a critical approach, the authors of Chapter 10 reflect on the possible definitions of "Europe". Etienne Toureille, Antoine Laporte and Claude Grasland show that the concern for its limits is part of a general tendency to divide up the world into regions, which is tinged with an already long-standing – and very European – concern about a possible decline of Europe.

– In order to leave the framework of classical regional geography, the same authors continue their reflection in Chapter 11 by presenting three empirical illustrations of "Europe": the position of European states in diplomatic networks; the membership of Turkey, a state in an interface position, to different supranational entities; the regional theme in French press articles. The latter is a methodology of great significance to follow the development of regional representations and therefore the possible passage from a simple *level* to an appropriate *territory*.

– Chapter 12 is dedicated to the Arctic. In this chapter, Frédéric Lasserre and Camille Escudé invite the reader to observe a macro-region in the making. Through the cooperation between riparian states, the action of regional institutions and the various interactions between the local, national, global and macro-regional levels, this framework plays an increasingly important role politically. In a space where regional issues have become clear to all, can regionalism precede the regionalization of trade?

– In Chapter 13, Christian Girault analyzes the role of NAFTA in the structuring of North America, which increasingly includes the Caribbean and Central American peripheries. This vast functional whole, devoid of powerful institutional governance, abused by various actors and challenged by former President Trump, ultimately proves to be solid.

– In Chapter 14, Sébastien Velut offers a critical approach to Latin America. Although the term has become part of everyday language and the discourse of international institutions, he shows that Latin America is hardly integrated and analyzes the obstacles to regional integration in this part of the world.

– In Chapter 15, Thierry Sanjuan and Karine Henriot recall that China's emergence is based, among other things, on a program of hegemonic domination of its neighbors. It is, in fact, a powerful factor of macro-regional integration. Beijing's action underwent reorientations in the 2000s: it now involves the construction of security cooperation networks in East and Southeast Asia. This regional undertaking can be seen as a basis for China to better position itself as a global actor.

– In Chapter 16, Nathalie Fau presents the situation in Southeast Asia. She shows the complex role of both Chinese hegemony and American power. Another difficulty with regional integration is the shift from functional integration, driven by transnational firms, to integration that is more formalized through free trade agreements, but complicated by the rise of authoritarian regimes in the region. In this context, ASEAN is having difficulty playing a leading role. Therefore, the author questions the capacity of civil societies to feed regional dynamics.

The concluding section summarizes the theoretical and empirical issues highlighted in this book and outlines what could be, in the future, a more visible francophone contribution to the comparison of regional integrations in the world.

References

Bello, W. (2021). Démondialisation, un paradigme à revendiquer. In *Démondialisation ?*, Polet, F. (ed.). Éditions Syllepses, Louvain-la-Neuve.

Latouche, S. (2019). *La décroissance*. PUF, Paris.

Sapir, J. (2010). *Le protectionnisme*. PUF, Paris.

1

From Regional Geography to the Geography of Regionalization

Pierre BECKOUCHE[1] **and Yann RICHARD**[2]
[1] *UMR Ladyss, Université Paris 1 Panthéon-Sorbonne, France*
[2] *UMR Prodig, Université Paris 1 Panthéon-Sorbonne, France*

1.1. Introduction

In the 19th century, academic geography emerged in Europe, particularly in France. In the wake of the work of Paul Vidal de la Blache, French geographers paid attention to sub-state regional divisions, in a monographic approach. However, from the beginning of the 20th century and during the interwar period, some observed what was already called globalization. They proposed a regionalization of the world, based on the geographical distribution of power centers and the observation of exchanges. After the war and until the 1970s, the themes of the regionalization of the world and regional integration on a macroscale (the construction of large multi-state groups) reappeared in France in a scattered manner. With their expertise on the notion of region, geographers, using their tools, took possession of this theme again, based in part on more advanced work from other disciplines.

Regionalization of the World,
coordinated by Pierre BECKOUCHE and Yann RICHARD. © ISTE Ltd 2024.

1.2. 19th and 20th centuries: the hesitant paths of regionalization between intra-state regions and continents

1.2.1. *The legacy of the 18th century*

To regionalize is to divide the surface of the Earth in order to classify knowledge about the known space. With the improvement of cartography and statistics, in the 18th century, regional descriptions became more precise: based on the political-administrative divisions, in states with an administration that produced figures. Whether on a global or national scale, the starting point was a naturalistic or even deterministic division. The emphasis was on climates, mountains, rivers and seas. Some works were based on watersheds (Philippe Buache and then Maxime Auguste Denaix), on altitude and climate (Abbé Giraud-Soulavie) and on geological outcrops (Omalius d'Halloy at the beginning of the 19th century, who planned to produce a rational division of the globe based on geology) (Claval 1999). This division played a central role in the French-speaking geographical vision in the 19th century, in the context of the exploration of the interior of the continents, the improvement of cartographic coverage and the census of reliefs and landscapes.

1.2.2. *Vidal and the vidalians: lifestyles, cities and circulation*

At the end of the 19th century, regional geography was defined in France by Paul Vidal de la Blache, who was influenced by German authors. Carl Ritter formalized the idea of regional differentiation by dividing the world according to a hierarchical model: the world, the continents and the parts of continents (mainly based on the observation of relief and climate) themselves divided into smaller regions. He divided Europe into regions according to several criteria: topography, atmospheric and hydrological data, population distribution, location of resources and distribution of fauna and flora. For him, the regions owe their personality to the particular relations which are established between the human societies and nature; their boundaries do not have any relationship with the political network.

Vidal de la Blache in turn laid the foundations of a regional approach that used concepts such as the "physiognomy" of regions, "ways of life", environments and landscapes, and that would mark the French school of geography until the 1940s. His region is a homogeneous whole defined by the singular interaction between a natural environment and the action of a social group. He later integrated other elements into his thinking: the role of cities (he already spoke of "nodal regions"), the exchange of goods and circulation in general. This prefigured the idea of functional regions and led him to reduce the importance of the principle of homogeneity when defining regional boundaries.

Vidal de la Blache contributed to the necessity of a monographic approach, but at the same time, following in the footsteps of F. Ratzel in Germany, posed some important questions about the regional fact at all scales. Are the combinations described unique or recurrent? Can we identify major types of regions from the examples described? What relationships are established between the identified regions? What role does regional differentiation play in the construction of the state? Not all the promises of Vidal de la Blache's work were fulfilled. Often written by his students, the great regional theses of the interwar period have some weaknesses. They were oriented towards the agricultural world, neglecting urbanization and industrialization; they forgot the new countries and the developing world; they favored small- or medium-sized territories, whereas long-distance exchanges existed and were increasing; the French regional approach remained at the intrastate level (see Vidal de la Blache (1903)), in a context that was nonetheless marked by internationalization.

Continents, regions and divisions of world space at the turn of the 19th and 20th centuries: Universal Geographies

Universal geographies (*Géographies Universelles*) reflected the orientations of French regional geography at that time. The criteria for regionalization were never explicit, but they were obviously influenced by the continental divisions inherited from the tradition since at least the Renaissance and taken at face value (Grataloup 2009). The divisions could have also been influenced by editorial contingencies, such as the desire to publish volumes of equivalent size. Between the 19th century and the middle of the 20th century, the most famous universal geographies were those of Conrad Malte-Brun (*Précis de géographie universelle*, then *Géographie universelle ancienne et moderne, mathématique, physique, statistique, politique et historique des cinq parties du monde*), Elisée Reclus (*Nouvelle géographie universelle : la terre des hommes*), Paul Vidal de la Blache and Lucien Gallois (*Géographie universelle*).

In his *Précis de géographie universel*, published between 1810 and 1829, Malte-Brun exposes an encyclopedic knowledge and his taste for a geography of discovery, a descriptive geography that does not propose any method of regionalization. He combines a continental division of the world with sub-continental regional divisions and a description of states and certain sub-state geographical entities (cantons, cities, provinces, etc.), without justifying his choices. In the *Nouvelle géographie universelle*, published in 1876 and 1894, Reclus does not propose a regionalization method of the world either, even though he questions what we would today call discontinuities, for example, within the Eurasian group (Bruneau 2018). His divisions were made a priori and its volumes were distributed by continental masses. Europe and Asia each have four volumes, Africa has five and Oceania has one. In each volume, he describes nested geographical entities: pieces of continents, states, pieces of states, river basins, etc. For example, Russia is divided between a European volume and a volume devoted to "Russian Asia", which surprises today's reader. The Mediterranean disappears as a region, split between "Southern Europe", "Northern Africa" and "Anterior Asia". It is not surprising that Reclus began his description of the limits of Europe by referring to an antiquity that is more than 2000 years old: "From their first expeditions of war or trade, the inhabitants of the eastern shores of the Mediterranean had to learn to distinguish the three continents that meet there" (*Nouvelle géographie universelle*, Volume 1, *L'Europe méridionale*, p. 9, author's translation). In the *Géographie universelle* by Vidal de la Blache and Gallois, the continental division of the world also prevails. The authors describe geographical entities of all sizes, according to a sequence whose meaning is not obvious. Olivier Orain recalls, for example, that the tome written by Raoul Blanchard on Western Asia "is divided into seven regional chapters, which draw an itinerary in the form of a spiral (Caucasia, separately; then Asia Minor, Armenia [in the broad sense], Iran, Arabia, Syria, and finally Mesopotamia)" (Orain 2000, author's translation). Each chapter is divided into sub-regions. Frequently, these sub-regions are in turn subdivided, either thematically, regionally or in a hybrid manner. At the end of the 20th century, *Géographie universelle*, edited by Roger Brunet and Olivier Dollfus, paid more attention to globalization. But the distribution of volumes remains influenced by the classical continental division of the world.

1.3. 20th century: an early but timid emergence of the world

1.3.1. *The world as a chessboard*

Although French geographers were marked by the Vidalian tradition in the first half of the 20th century, some were already paying attention to the effects of the internationalization of exchanges regarding the organization of space. Even before World War II, they observed the improvement of means of communication and the extension of colonial empires. As Jean-Baptiste Arrault showed, they concluded that the world had already existed in the 19th century, that it was a scale of geography and an object whose regional organization had to be understood. Although this research remained secondary in the production of French geographers at that time, it gave rise to regular publications that were sometimes strikingly clear-sighted.

René Crozet, Albert Demangeon, André Siegfried and Maurice Zimmermann were interested in the opening of the great international canals (Panama and Suez). They wanted to show that it divided the world economy into new economic areas (Arrault 2007). The opening of the Suez strengthened the interaction between the center (industrialized Europe) and a vast periphery (the Orient, India and Australia). The canal was perceived as a European route (Vergez-Tricom 1920) and Panama as a route that reinforced what we would today call American regional integration (Zimmermann 1916). In an approach that could be described as geo-economic, Siegfried saw the "respective zones of Suez and Panama" as a competition between the main economic powers of the inter-war period (Japan and the United States), at the expense of Europe (Siegfried 1940).

Politically, the world was perceived as a chessboard: the emerging geopolitics now influenced geography. Outside of France, Mahan (1890), Ratzel (1903) and Mackinder (1904) questioned the factors of power: the balance of power was between land and sea powers. Mackinder (2004) proposed a formal regionalization of the world, which would become a milestone. Intended for policy makers, their work mixed scientific posture, ideological bias and political prescriptions. This is a thread that authors such as Haushofer (1931) and Schmitt (2003 [originally published in 1950]) then followed, with an even more ideological content. These approaches were denounced by French geographers who, in a form of critical geopolitics before its time, saw in German *Geopolitik* a propagandistic discourse at the antipodes of science (Demangeon 1932a; Ancel 1936a). As a counterpoint, they proposed their geopolitics of areas of influence. Thus, Vidal de la Blache described the clash of Russian and British imperialism in central Asia (Vidal de la Blache 1889) and Zimmermann the ambitions of Japan "in the East of Asia on a vast portion of the world chessboard" (Zimmermann 1908, author's translation).

1.3.2. *The inter-war period: decline of Europe, the European region, Triad*

In *Le déclin de l'Europe*, Demangeon distinguishes what would later be called the Triad: "the division is between Europe, North America and the Japanese Archipelago [...] There are now several centers of high humanity instead of one. Since the great discoveries, the world had become Europeanized; under the influence of continents and peoples younger in progress, it tends to become regionalized" (Demangeon 1920a, author's translation). Even before World War I, Vidal de la Blache described (in his *Principes de géographie humaine*) what we now call the world system, structured by these three poles and their exchanges.

The Pacific became a region of the world for geographers who were interested in the intensification of relations between its shores, the rise in power of the United States and Japanese migratory pressure (Arrault 2007). For Zimmermann (1897, 1899), the Pacific was the space where we had to gain a foothold if we hoped to count in international competition; the "Pacific question" was linked to the increase in links between its shores. In his course at the Collège de France, Brunhes was interested in the Pacific, which he saw as a space as such. Finally, Demangeon (1920b) spoke of the advent of the Pacific, which "tends to become a great economic unit".

French geographers became aware of the need for Europe to unite in order to resist this new global competition. A European regional problem was born. Demangeon (1913) said that Western Europe was a "commercial city" in the face of America. So, the union of Europe was envisaged in academic debates. With an impressive foresight, Demangeon (1929) believed that the European idea offered a perspective for dealing with the economic crisis by specializing "in quality production" and developing international agreements through industrial cartels. Like Jacques Ancel, he criticized the political fragmentation of Europe as a major cause of economic problems and promoted "regional agreements". Echoing these works, Siegfried (1929) gave a history of European sentiment, which had grown stronger since the war. He also considered the possible rapprochement of European states "as a prospect of salvation for a group of powers that are experiencing decline" (Siegfried 1935). For all these geographers, although it was no longer the master, Europe was then represented as a region of the world; it was therefore necessary to think about it and organize it. In the eyes of Vidal de la Blache and Siegfried, it was a civilization that, before being a continent, shared values: humanism, individualism, invention and freedom of the mind. This revived the classical discussion on its geographical boundaries and inaugurated a vein that structured the debates of the 20th century on a cultural definition of the regions. Demangeon hesitated: he excluded Russia from the political point of view (with the Bolshevik regime being incompatible with democracy), but considered that its presence in Europe was necessary for the economic balance of the continent. For Ancel, on the other hand, the physical and human geography of Russia distanced it from Europe (Demangeon 1932b; Ancel 1936b; Arrault 2007).

1.3.3. *After World War II: global regions at last?*

After World War II, regional geography went beyond the first Vidalian tradition, i.e. the relations considered too simple between human groups and environments, the rural tropism, and especially the monographic approach. From the 1960s onwards, regional monographs took a back seat. The notion of spatial organization became central; it encompassed what was happening within the region *and* the relations with other regions. Some geographers refer to this as the "system region" (Dumolard 1975). In the wake of Walter Christaller's economic theories and the theoretical contributions of economists such as François Perroux, the region was considered less in its homogeneity and more through the idea of functional integration. Regionalization then consisted of observing the spatial limits of economic channels and identifying nodes (cities) that dominated surfaces. We find this idea in Juillard (1962). The young regional science seeks laws of geographic space and studies the forces that organize it according to repeating patterns (Isard 1975). To understand regions, it is necessary to analyze networks, economies of scale, inter-scale relationships, production relationships and center–periphery relationships. This is what Marxist-inspired radical geography does. Moreover, spatially sensitive economists, such as Lipietz (1977), push geographers to seek principles of spatial division of labor valid at the intra-national level, as well as the international level.

From the end of World War II to the 1980s, however, regional francophone geographies of the world remained scattered. They were the work of a few prominent authors who developed different approaches, without creating a school of thought, with the publication of numerous works or articles devoted to this or that country (geography of Japan, France, Brazil, etc.) or to this or that part of the world[1]. Despite the path opened up by Fernand Braudel's Mediterranean, reflection on regionalism and regional integration on a macroscale remained marginal in France, both among geographers (including Dollfus, who was nevertheless receptive to Braudel's work) and historians. Braudel's intuitions were instead extended by economists and sociologists (Wallerstein, Arrighi and Amin). According to Grataloup (2018), this discretion is understandable if we place it in the intellectual context of post-modernity, which was not very favorable to the historical study of large regional areas. Talking about the East, Chinese, African, Amerindian and Oceanian worlds was difficult in the face of the rise of postcolonial criticism, the deconstruction of western subjectivity and the representations associated with it. The global scale was taking over the macro-regional scale; with the rise of global history, little attention was paid to regional divisions.

1. Braudel's "civilizational areas" are related to what may be called a multi-monographic approach, in which the deep invariants of a region prevail over its external interactions. The idea of civilizational area – reappropriated and reinterpreted – is found in Pierre Gourou's work, who explores the Braudelian idea of the civilization of rice.

This is why works on large regional groups remained rare among geographers. Claval (1968), however, is an exception. He took up the idea of international trade organized around three major poles. But his analysis of "large territorial groupings" was mainly economic and left aside the work of political scientists specializing in international relations who were working on theories of regional integration at the time (Haas in particular). We should also mention the geographers specializing in the Soviet Union, who necessarily took a macro-regional perspective given the size of the area they studied: a space of more than 22 million square kilometers, made up of 15 republics (George 1962; Blanc and Chambre 1971; Maurel 1982; Radvanyi 1982; Carrière 1984). They showed how planning allowed the establishment of social, economic and political links between these republics, which made the Soviet Union function as an integrated whole for several decades. However, the Soviet case remained ambiguous: should we consider that the USSR was a state, or a group of neighboring states engaged in regional integration?

1.3.4. *The 1980s: the world as a system*

Systemic approaches to the world revived francophone geography from the 1980s onwards. In his research on spatial justice, Reynaud (1981) was influenced by the work of Amin (1973) on dependence, and analyzed the center–periphery model at all scales. This led him to propose a regionalization of the world based on the relations between centers and peripheries, following the example of Immanuel Wallerstein who observed the dynamics of capitalism in the world system.

In his work on globalization, Dollfus (1984) further developed the idea of the "world system". The world is no longer just a framework, it is a geographical object per se, a system of interlocking and connected socio-economic systems, a kind of large multipolarized region. In the 1990s, he even proposed a regionalization of the world, but emphasized other entities in this regionalization, such as cities and what he called the "global megalopolitan archipelago" (Dollfus 1994). In his work, he showed above all that a few major state powers and a few hundred global economic and financial groups dominated and organized the world system, with "driving spaces", "polarized spaces" and "spaces in the process of destructuring".

1.4. The turn of the 20th/21st century: relations between geographers and other disciplines are almost one way

Three changes mark the francophone regional geography at the turn of the two centuries.

1.4.1. *Economics reformulates the regional equation*

Geographers first take into account the rapid pace of globalization and the intensification of the technological content of production. These aspects become decisive in regional analysis, because the acceleration of the mobility revolution, particularly that of information, make physical criteria less relevant for dividing up regions. This leads them to increasingly take into account the work of economists on regional integration. Based on the theory of optimal currency zones (Robert Mundell) and the stages of economic regionalism (Bela Balassa), they broaden their work on regional integration by comparing intraregional dynamics with the growth of liberalized trade on a global scale. The pioneering analyses of the economists of the 1960s and 1970s on European integration are continued by a systematic reading of regional economic integration in the world based on three indicators:

– The convergence or divergence of national economies belonging to the same regional grouping: on this theme, economists propose analyses that geographers would not disavow (Hugon 2003; Deblock and Regnault 2006). With the new economic geography, they launch a challenge that stimulates the latter (Krugman 2000).

– The geography of flows, in order to see whether exchanges are increasingly, or less and less, intraregional (Chaponnière 2004).

– The nominal objectives of regional trade agreements, where a region is considered to be in the process of economic integration if its specific objectives are translated into reality (Baldwin 1997). By distinguishing between superficial regional integration (simple free trade objectives) and deep integration (regionalization of value chains and shared standard objectives), economists provide essential concepts useful for understanding regions.

1.4.2. *Trendy regionalism and multidisciplinary necessity*

The second change in the landscape comes from international law and political scientists who have been studying the trend, since the mid-1990s, for regional intergovernmental agreements, including Regional Trade Agreements (Dieter 2006; De Lombaerde et al. 2010), or post-Soviet agreements that aim to maintain Russia's area of influence over its peripheries by building new regional organizations (Commonwealth of Independent States, Eurasian Economic Union, Collective Security Treaty Organization). French-speaking geographers become accustomed to drawing on economic and political approaches, whether to understand regional integration or disintegration, or to compare the institutional regions drawn by the RTAs with the functional realities of the territories. This has led to a renewal of the analysis of world regions (Mareï and Richard 2018) conducted by geographers, often in multidisciplinary teams (Gemdev 1994; Azuelos et al. 2004; Taillard 2004; Radvanyi and Laruelle 2016).

In this inevitable interdisciplinarity, the contribution of geographers is multiform. We can mention five fields.

They analyze functional territories and, in particular, the constitution of international networks whose cross-border actors are migrants or informal traders rather than the firms and states that dominate economic or political analyses. In this field, geographers mobilize the work of sociologists working on informal networks (Tarrius 2002), co-creating with them an approach to regionalization "from below" (Vignal 2020).

They measure the regionalization of flows: to do so, they use various indicators (trade, investment, migration and mobility, tourism, ship movements, etc.) to map flows and networks (Grasland and Didelon 2006; Zdanowska 2020) and measure spatial interactions.

They describe *concrete* territories, in particular because they are used to dealing with interactions between all scales. In doing so, they make a useful contribution to the paradigm that has become established in the spatial reading of the international economy, that of an archipelago economy (Veltz 1996). They give importance to the proximity of kilometers (Dollfus 1994) in the face of economists and political scientists who often tend to reduce the actors to planetary-scale strategists. The approach of geographers is not necessarily statocentric; their analyses are not built on national bricks alone. They remind us of the importance of the long term, which establishes tribal solidarities here or commercial solidarities there, and which organizes international urban networks that cross historical periods. They recall the particular reasons why a given regional territory does not really verify the general economic laws of regional integration, or follow the guidelines and perimeter of an intergovernmental treaty as we would expect, which leads them to question the links between regionalism and regional integration (Van Hamme 2014). They compare representations and discourses (Beauguitte et al. 2012) with the realities on the ground. They analyze the territorial consequences of regionalism: concrete effects of the European Union's regional policy and cohesion policy (decrease or increase in territorial inequalities, realities of multi-level governance, etc.), the Infrastructure Integration Initiative of the South American region, corridors in ASEAN or in African regional organizations (Elissalde and Santamaria 2008; Baudelle and Jean 2009; Boulineau 2017; Santamaria 2017).

In particular, they analyze the functioning of borders, whose permeabilization is the basis of international regions. One of their contributions has been to show the change in the status of the *neighbor*: they used to be the hereditary rival for access to passageways and essential resources such as people, minerals, arable land or water. In the era of the knowledge economy, where the vast majority of resources are produced and no longer taken, the neighbor becomes the privileged partner for the co-production of resources and the cooperative expansion of market areas (Drevet 2008; Beckouche 2008; Beckouche et al. 2016). Their contribution is notable in the observation of border regions that become places of more or less in-depth cooperation, because regional integration consists precisely of reducing the barrier effects of political limits as much as possible. From this point of view, European cross-border and transnational cooperation is the focus of most research (Hamez 2001; Reitel and Wassenberg 2020), but French geographers are working on other parts of the world, notably Latin America (Medina 2009; Medina and Diallo 2020).

Finally, they take into account the environmental dimension of regionalization. Natural environments, which had somewhat disappeared from the geography of the *Trente Glorieuses*[2], made a strong comeback at the beginning of the 21st century. Through the notion of transnational public goods, environmental issues are even becoming essential objects of regional geography: common regional regulation against marine pollution, transboundary management of water resources and "hydrodiplomacy", coordinated management of natural disasters and crises, coordination in the field of nuclear safety or energy transition. But it is essentially as a *result* of the powerful anthropic marking of territories that the environment is becoming an issue of regional geography, instead of the role of the territorial determinant that the founders of the regional division had once given to natural environments. This avoids the excessive naturalization of regional groupings that led to their a priori definition (on the critical approach to regional divisions, see Grataloup (2011), Capdepuy (2012) and Didelon Loiseau (2013)), and better illustrates the consensus that sees every region as a social construction.

1.4.3. *Contributions and aporias of the civilizational paradigm*

The third change has come from sociologists and political scientists working on the notions of "civilization" and culture. In particular, today's regional geographers can hardly ignore the "cultural" paradigm propelled to the top of the agenda by Samuel Huntington in the 1990s. Despite the growing importance of the cultural theme in the contemporary regional equation, their joint work with anthropologists and sociologists, and their contribution to a better understanding of tribal realities

2. Thirty years following the end of World War II in France.

(Balanche 2012) in regions that westerners tended to analyze through their own cultural lenses too much, it is not clear that geographers have, on the whole, provided a consistently convincing response to the powerful reductionism of culture to religion, and of religion to simplism that has made Huntington's followers so successful. Critical studies of the civilizational division of the world by geographers remain rare (Didelon Loiseau and Richard 2020).

Three final points conclude this quick overview of the recent francophone regional geography. These works are, on the whole, not sufficiently known by other disciplines. Perhaps, the influence of monographic approaches is still too great. And, although most of these works use convergent approaches, it must be recognized that the methodological coordination of geographers is still perfectible if the rise in generality of the new regional geographies is to become readable by the other disciplines. Finally, there is still no consensus on the meaning of words, which makes comparisons difficult (Richard 2014; Mareï and Richard 2020). Nevertheless, the contribution of geographers is notable. This book, we hope, is a testament to that.

1.5. References

Amin, S. (1973). *Le développement inégal*. Les Editions de Minuit, Paris.

Ancel, J. (1936a). *Géopolitique*. Delagrave, Paris.

Ancel, J. (1936b). *Affaires étrangères : aide-mémoire de la politique étrangère française (1789–1936)*. Delagrave, Paris.

Arrault, J.-B. (2007). Penser à l'échelle du Monde. Histoire conceptuelle de la mondialisation en géographie (fin du XIXe siècle/entre-deux-guerres). Geography PhD Thesis, Université Paris 1 Panthéon-Sorbonne.

Azuelos, M., Cosio-Zavala, M.-E., Lacroix, J.-M. (2004). *Intégration dans les Amériques. Dix ans d'Alena*. Presses Sorbonne Nouvelle, Paris.

Balanche, F. (2012). *Atlas du Proche-Orient arabe*. Presses Sorbonne Université, Paris.

Baldwin, R. (1997). The causes of regionalism. *The World Economy*, 20(7), 865–888.

Baudelle, G. and Jean, Y. (eds) (2009). *L'Europe. Aménager les territoires*. Armand Colin, Paris.

Beauguitte, L., Didelon, C., Grasland, C. (2012). The EuroBroadMap project: Visions of Europe around the world. *Politique européenne*, 37(2).

Beckouche, P. (2008). *Les régions Nord-Sud*. Belin, Paris.

Beckouche, P., Besnard, P., Pecout, H. (eds) (2016). *Atlas of Challenges and Opportunities in European Neighbourhoods*. Springer Open, Heidelberg.

Blanc, A. and Chambre, H. (1971). *L'URSS*. PUF, Paris.

Boulineau, E. (2017). Les macrorégions en Europe, une nouvelle régionalisation pour renforcer la cohésion territoriale. In *Les Régions et la politique de cohésion de l'Union européenne*, Baudelle, G. and Charles-Le Bihan, D. (eds). PUR, Rennes.

Bruneau, M. (2018). L'Eurasie, un impensé de la géographie : continent, empire, idéologie ou projet ? *L'Espace Géographique*, 47(1), 1–18.

Brunet, R. (eds) (1990–1996). *Géographie universelle*, 10 volumes. Hachette, Belin-Reclus, Paris.

Capdepuy, V. (2012). Un espace : l'Eufrasie. *M@ppemonde*, 104 [Online]. Available at: http://mappemonde.mgm.fr/num32/articles/art11403.

Carrière, P. (1984). *L'économie de l'URSS*. Masson, Paris.

Chaponnière, J.-R. (2004). L'électronique en Asie orientale, division internationale du travail et intensification des échanges. In *Intégrations régionales en Asie orientale*, Christian, T. (ed.). Les Indes savantes, Paris.

Claval, P. (1968). *Régions, nations, grands espaces. Géographie générale des ensembles territoriaux*. Marie-Thérèse Génin, Paris.

Claval, P. (1999). *Introduction à la géographie régionale*. Nathan, Paris.

De Lombaerde, P., Söderbaum, F., Van Langenhove, L., Baert, F. (2010). The problem of comparison in comparative regionalism. *Review of International Studies*, 36(3), 731–753.

Deblock, C. and Regnault, H. (2006). *Nord-Sud : la reconnexion périphérique*. Athéna, Montreal.

Demangeon, A. (1913). Les relations de la France du nord avec l'Amérique. *Annales de Géographie*, 22(123), 227–244.

Demangeon, A. (1920a). *Le déclin de l'Europe*. Payot, Paris.

Demangeon, A. (1920b). Les relations économiques de l'Australasie. *Annales de Géographie*, 157, 75–78.

Demangeon, A. (1929). Les aspects actuels de l'économie internationale. *Annales de Géographie*, 211, 10–25.

Demangeon, A. (1932a). Géographie politique. *Annales de géographie*, 41(229), 22–31.

Demangeon, A. (1932b). Les conditions géographiques d'une union européenne. Fédération européenne ou ententes régionales ? *Annales d'histoire économique et sociale*, 17, 433–451.

Didelon Loiseau, C. (2013). Le Monde comme territoire, pour une approche renouvelée du Monde en géographie. HDR, Université de Rouen.

Didelon Loiseau, C. and Richard, Y. (2020). La géographie à la recherche des civilisations de Huntington. Analyse des représentations du monde. *Bulletin de la Société Géographique de Liège*, 74, 5–22.

Dieter, H. (ed.) (2006). Report on east Asian integration opportunities and obstacles for enhanced economic cooperation. *Notre Europe*, Studies & Research, 47.

Dollfus, O. (1984). Le système Monde. Proposition pour une étude de géographie. *Actes du Géopoint 1984, Systèmes et localisations,* Groupe Dupont, Université d'Avignon, 231–240.

Dollfus, O. (1994). *L'Espace-Monde*. Economica, Paris.

Drevet, J.-F. (2008). L'Union européenne et ses périphéries : faisabilité du scénario des dominos. *Strates*, 15, 25–36.

Dumolard, P. (1975). Région et régionalisation. *L'Espace Géographique*, 4(2), 93–111.

Élissalde, B. and Santamaria, F. (2008). *Lexique de l'aménagement du territoire européen*. Lavoisier, Paris.

Gemdev (1994). *L'Intégration régionale dans le Monde*. Karthala, Paris.

George, P. (1962). *L'URSS*. PUF, Paris.

Grasland, C. and Didelon, C. (2006). Europe in the World. Final report, ESPON Project 3.4.1, UMS Riate, Paris.

Grataloup, C. (2009). *L'invention des continents : comment l'Europe a découpé le monde*. Larousse, Paris.

Grataloup, C. (2011). *Faut-il penser autrement l'histoire du monde ?* Armand Colin, Paris.

Grataloup, C. (2018). Histoire. In *Dictionnaire de la régionalisation du monde*, Mareï, N. and Richard, Y. (eds). Atlande, Neuilly.

Hamez, G. (2001). Du transfrontalier au transnational : approche géographique : l'exemple de la frontière franco-belge. Geography PhD Thesis, Université Paris 1.

Hauser, H. (1926). Qu'est-ce que l'Europe ? *Le Monde nouveau : revue mensuelle internationale*, 6–7, 681–688.

Haushofer, K. (1931). *Geopolitik der Pan-Ideen*. Zentral-Verlag, Berlin.

Hugon, P. (2003). *Les économies en développement à l'heure de la régionalisation*. Karthala, Paris.

Isard, W. (1975). *Introduction to Regional Science*. Prentice Hall, Englewood Cliffs.

Juillard, E. (1962). La région : essai de définition. *Annales de géographie*, 386, 483–499.

Krugman, P. (2000). Where is the world in the new economic geography? In *Oxford Handbook of Economic Geography*, Clark, G., Feldman, M., Gertler, M. (eds). Oxford University Press.

Lipietz, A. (1977). *Le capital et son espace*. Maspero, Paris.

Mackinder, H.J. (1904). The geographical pivot of history. *The Geographical Journal*, 170(4), 298–321.

Mahan, A. (1890). *The Influence of Sea Power upon History, 1660–1783*. Little, Brown & Company, New York.

Malte-Brun, K. (1810–1829). *Précis de géographie universelle*, 8 volumes. Eugène et Victor Penaud Frères, Paris.

Mareï, N. and Richard, Y. (2018). *Dictionnaire de la régionalisation du monde*. Atlande, Neuilly.

Mareï, N. and Richard, Y. (2020). Global regionalization processes and (macro)-regional integration. Favor of generalization in geography. *Belgeo*, 4.

Maurel, M.-C. (1982). *Territoire et stratégies soviétiques*. Economica, Paris.

Medina, L. (2009). Les frontières de l'isthme centraméricain, de marges symboliques à des espaces en construction. *Espaces et sociétés*, 138, 35–50.

Medina, L. and Diallo, M. (2020). Les coopérations transfrontalières comme outils d'intégration régionale : analyse croisée dans les Suds (Amérique centrale, Afrique de l'Ouest). *Belgeo*, 4 [Online]. Available at: http://journals.openedition.org/belgeo/43693.

Orain, O. (2000). Les "postvidaliens" et le plain-pied du monde. Pour une histoire de la géographie au risque de l'écriture. In *Logiques de l'espace, esprit des lieux. Géographies à Cerisy*, Lévy, J. and Lussault, M. (eds). Belin, Paris.

Radvanyi, J. (1982). *L'URSS : régions et nations*. Masson, Paris.

Radvanyi, J. and Laruelle, M. (2016). *La Russie, entre peurs et défis*. Armand Colin, Paris.

Ratzel, F. (1903). *Politische Geographie oder die Geographie der Staaten, des Verkehres und des Krieges*, 2nd edition. Druck und Verlag von R. Oldenburg, Munchen, Berlin.

Reclus, E. (1876–1894). *Nouvelle géographie universelle*, 19 volumes. Hachette, Paris.

Reitel, B. and Wassenberg, B. (eds) (2020). *Critical Dictionary on Borders, Cross-Border Cooperation and European Integration*. Peter Lang, Brussels, Berlin.

Reynaud, A. (1981). *Société, espace, justice*. PUF, Paris.

Richard, Y. (2014). Intégration régionale, régionalisation, régionalisme – Les mots et les choses. *Confins*, 20.

Santamaria, F. (2017). L'UE en quête d'aménagement de son espace. *L'Information géographique*, 81(4), 129–151.

Schmitt, C. (2003). *The Nomos of the Earth in the International Law of the Jus Publicum Europaeum*. Telos Press, New York.

Siegfried, A. (1929). European reactions to American tariff proposals. *Foreign Affairs*, 1, 13–19.

Siegfried, A. (1935). *La crise de l'Europe*. Calmam-Lévy, Paris.

Siegfried, A. (1940). *Suez, Panama et les routes maritimes mondiales*. Armand Colin, Paris.

Taillard, C. (2004). *Intégrations régionales en Asie orientale : NORAO*. Les Indes savantes, Paris.

Tarrius, A. (2002). *La Mondialisation par le bas : les Nouveaux Nomades de l'économie souterraine*. Balland, Paris.

Van Hamme, G. (2014). Régionalisme, intégration régionale et mondialisation : quelques éclairages sur un triangle complexe. In *La régionalisation du monde. Construction territoriale et articulation global/local*, Gana, A. and Richard, Y. (eds). Karthala, Paris.

Veltz, P. (1996). *Mondialisation, villes et territoires. L'économie d'archipel*. PUF, Paris.

Vergez-Tricom, G. (1920). Le canal de Panama. *Annales de Géographie*, 161, 384–387.

Vidal de la Blache, P. (1889). *Etats et nations de l'Europe. Autour de la France*. Delagrave, Paris.

Vidal de la Blache, P. (1903). *Tableau géographique de la France*. Hachette, Paris.

Vignal, L. (2020). Mondialisation, régionalisation, conflits : comprendre le monde à partir d'un Moyen-Orient en ébullition. HDR in Geography, Université Lyon-3.

Zdanowska, N. (2020). L'Europe centrale et orientale au sein des réseaux mondialisés de l'économie-monde : quelle intégration interne ? *Belgeo*, 4 [Online]. Available at: http://journals.openedition.org/belgeo/43462.

Zimmermann, M. (1897). L'annexion de l'archipel Hawaï aux Etats-Unis. L'importance internationale du Pacifique. *Annales de Géographie*, 30.

Zimmermann, M. (1899). La cession des Carolines à l'Allemagne. *Annales de Géographie*, 41.

Zimmermann, M. (1908). Le développement du commerce et de la marine marchande du Japon. *Annales de Géographie*, 91, 88–92.

Zimmermann, M. (1916). La première année du trafic du canal de Panama. *Annales de Géographie*, 134, 157–158.

2

The Regionalization of Migration

Camille SCHMOLL
EHESS/Géographie-cités, Institut Convergence Migrations, Paris, France

2.1. Introduction

In recent decades, international migration has increased considerably: UN data indicate that there were 153 million migrants in the world in 1990, rising to 281 million in 2021[1]. Beyond this numerical observation, most contemporary works insist on an essential characteristic of these flows, namely their globalization (de Haas et al. 2014). This same observation can be found in the public policy sphere: the notion of globalization is at the heart of the recent Global Compact for Safe Migration adopted by the UN in Marrakech in 2018.

This observation of a globalization of migration is doubly problematic. On the one hand, migration on a global scale is not a recent phenomenon, whether we think of major phases of displacement such as the transatlantic slave trade or the great transcontinental migrations of Europeans to the Americas. On the other hand, contemporary migration is – and has remained over the years – a predominantly regionalized phenomenon. It is to this latter aspect of contemporary migration dynamics, still largely underestimated in the scientific literature, that this chapter is devoted: what are the main characteristics of the regionalization of migration? Does it affect all regions equally? Do these dynamics also concern forced migration

1. That is, 2.9% and 3.6% of the world population, respectively.

Regionalization of the World,
coordinated by Pierre BECKOUCHE and Yann RICHARD.

(section 2.2)? After answering these questions, we will see what concepts geographers use to describe the spaces resulting from this regionalization. We will see that migratory circulation and transnational mobilities, two phenomena that have greatly inspired the geography of migration in recent decades, have contributed to weaving dense and dynamic relational forms that reinforce the regional dimension of migration (section 2.3). The third part of the chapter will address the political dimension of these regional groupings (section 2.4).

2.2. Regionalization: main characteristics

2.2.1. *A global trend...*

The phenomenon most described in the literature on international migration is therefore the globalization of migration: widening of the migratory fields; diversification of the groups present in the destination countries; diversification of the countries of departure and settlement, i.e. all countries of the world are affected by migration and, more often than not, experience both immigration and significant emigration. This last aspect is known and widely studied for certain countries such as Morocco, Mexico and Turkey, but it also concerns many European, South American, Asian and African countries. Nigeria, for example, attracts many people from the region (Mali, Benin, Ghana, Togo) and is also a country of departure to North America, Europe or South Africa.

However, the regionalization of migration remains an important feature of international migration and even tends to be reinforced in certain contexts. Globally, migration is organized into large regions within which multidirectional flows are observed. In 2020, nearly half of all international migrants resided in their region of origin. The regionalization of migration concerns both so-called "voluntary" and "forced" migration, although this dichotomous distinction has been the subject of much criticism (Schmoll 2020) and concerns all major regional groupings, with significant variations.

At first glance, the idea of a regionalization of migration may come as a surprise, as our stereotypes tend to point towards migration from the least wealthy countries to the most developed. In France, such a vision was notably fueled by an essay published in 2018 that met with great success: *The Scramble for Europe* by journalist Stephen Smith (2018). But many critical works have deconstructed this unidirectional vision. Several authors have shown how the usual mapping of international migration in Europe at the initiative of the media, large international organizations – such as the International Organization for Migration (IOM) – or even certain European agencies such as Frontex, had contributed to developing a biased vision – Western-centric or Eurocentric – of migration routes pushing economic migration from the South to the North (Lambert 2015; Bahoken 2021; see also Van Houtum et al. 2020). Behind this unidirectional view is the mistaken idea of a spillover of migration from poorer to richer areas when most migration takes place close to and within the same type of region in terms of levels of economic wealth (Héran 2018).

The regionalization of migration has several causes: it is linked to the effects of cultural and linguistic proximity between countries, and to the possibility (or at least the desire) of a close return, which is not necessarily definitive, moreover. It is often forgotten, but migrants usually leave behind a family, possessions and loved ones; the possibility of frequently returning to the place of origin can contribute greatly to the attractiveness of less distant destinations. Secondly, the phenomenon is fed by the effects of the migration chain, i.e. the presence of communities already established in the closest areas.

The regionalization of migration can take different forms. When it is largely concentrated in a single adjacent or nearby country (Greece–Albania, Libya–Egypt or Libya–Tunisia, Spain–Morocco, France–Algeria, Mexico–United States), we can speak of migratory pairs linked to proximity logics (Dumont 2009). But these coupling logics are partly challenged today, and regional migrations are more multipolarized.

2.2.2. *...particularly strong in Europe and Sub-Saharan Africa*

According to UN data (see Figure 2.1), Europe is the region with the largest share of intraregional migration in its total migration, with 70% of all European-born migrants residing in another European country. After Europe, Sub-Saharan Africa has the largest share of intraregional migration out of its total migration (63%), in complete contradiction to the common representation that most African migration is directed to Europe (Flahaux and De Haas 2016). Among the most significant migrations within the African continent are those from Burkina Faso to Côte d'Ivoire (1,400,000 people in 2019) and from Côte d'Ivoire to Burkina Faso (550,000 people in 2019); from Mozambique to South Africa (700,000 people), from Somalia to Ethiopia (500,000 people) and Kenya (450,000 people). These multidirectional migrations sometimes lead to reciprocal population exchanges (between Burkina and Côte d'Ivoire, for example) and help to organize Africa into vast regional circulation areas.

Unlike Europe and Africa, Central and South Asia is characterized by a large share of its diaspora residing outside the region (78%), followed by North America (75%) and Latin America and the Caribbean (74%). That said, these distinctions between regional groupings can be qualified, as they are also based on the regional grouping proposed by the United Nations: some interregional migration actually takes place within a certain proximity. This is the case for migration from India, Bangladesh, Pakistan, Nepal and Sri Lanka to the Gulf countries. This South Asian migration to the Gulf countries tripled between 2000 and 2020, from 7 to 20 million migrants (UN 2020; Valenta 2020).

2.2.3. *Contrasting developments*

As Figure 2.1 shows, the regionalization of migration has increased in recent years in several regional groupings, such as Latin America and the Caribbean, North Africa and the Middle East, and Europe. There are several reasons for this increase. In Europe, European integration and the entry of new countries into the common space of free movement explain it. In other regions, one of the explanations lies in the imposition of increasingly high barriers to migration from the global South (low-income countries) to the global North (developed countries). This mechanism of "frontierization" of the richest states (United States, European Union) towards the states of the global South was reinforced after the economic crisis of 2008 and contributed – at first – to strengthening intraregional migration within the countries of the global South. The same happened with the Covid-19 health crisis, which contributed to discouraging or even hindering international migration, especially over long distances. However, it should be noted that the policy of outsourcing migration control by the states of the global North can also have the opposite effect, namely to strongly limit intraregional migration. Loren Landau's work has shown how migration control imposed by the European Union on African states has contributed to forms of immobilization or slowing of international mobilities (Landau 2019).

In Latin America, the share of intraregional migration has increased steadily in recent years. This is largely due to the declining attractiveness of Spain in the years following the economic recession of 2008–2011 and, above all, to the significance of emigration from Venezuela starting in 2013. While Venezuela was, along with Argentina, the most significant immigration state in Latin America, it has suffered a massive exodus in the last eight years: by 2019, the Venezuelan population had reached more than 1 million in Colombia and 600,000 people in Peru. It is known, moreover, that these data greatly underestimate Venezuelan emigration, which in 2019 involved a total of 3.7 million people (Cerrutti 2020).

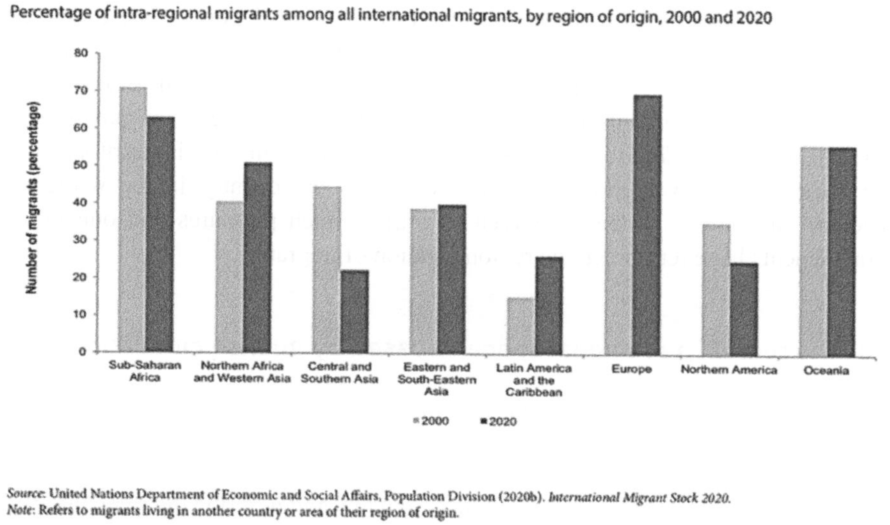

Figure 2.1. *Percentage of intraregional migration among all international migration, by region of origin, 2000/2020 (source: from UN DESA 2020). For a color version of this figure, see www.iste.co.uk/beckouche/regionalization.zip*

2.2.4. *The share of forced migration*

As the Venezuelan example illustrates, forced migration has notably increased in recent years. The top three groups of forced migrants in 2019 were Syrians (6.7 million people), Venezuelans (4 million) and Afghans (2.6 million). All three groups were concentrated in their region of departure. Thus, Syrian refugees are distributed among Turkey, Lebanon, Jordan and Iraq; Venezuelans among Colombia, Peru, Chile and Argentina; and Afghans mainly in Pakistan and Iran.

This has not generally led to distribution mechanisms from poorer states to richer ones. We might have imagined that the refugee resettlement programs set up by the most developed states (the United States, Canada and some European states) would have compensated for these proximity migrations, but this is not the case. These programs, which are extremely modest, involved a few tens of thousands of people in 2020 (only 34,383 people for resettlement programs).

On the other hand, the increase in displacement and forced migration has been accompanied by the delegation of control and reception – sometimes in camps or precarious settlement areas – as part of a vast policy of outsourcing the reception of refugees. This policy often contradicts the Geneva agreements and places a disproportionate burden on certain countries of the Global South. One example is the US agreements with Mexico under the Remain in Mexico program, implemented under Donald Trump. The program aimed to outsource the asylum procedures of Central American migrants to Mexico. In the European context, the pressure and agreements with Turkey to take care of exiles transiting through its territory are of the same nature (even though Turkey is already the country in the world that receives the largest number of forced migrants). Such pressures and outsourcing arrangements have reinforced the regionalization of migration.

2.3. The region: a relational space, crossed by intense circulation

2.3.1. *Naming and describing the regionalization of migration. From the field to the migration system*

What are the terms used by geographers to name and describe the regionalization of migration? Two expressions have marked the description of migration regionalization: "field" and "migration system". The term "migration field" refers to:

> Spaces traversed and structured by all the relatively stable and regular flows of migrants, whatever their origin or destination. This concept makes it possible for of all to better identify – well beyond the geography of the placed concerned – the specificity of the space practices by each group, at different levels (extended family, village group, ethnic-regional group, etc.), whatever the scale (local, national, regional, intercontinental). Thus, the migrant field of Moroccans, whose emigrants are established in a fairly large number of European countries and who are now settling in North America, is now wider and more open than that of Algerians or Tunisians, who remain concentrated in a small number of countries, including France (Simon 2002, p. 38).

Gildas Simon applies this notion to the relational space forged by Tunisian migrants between France and Tunisia. Michel Poinard also uses it to refer to the transnational migratory space of the Portuguese (Poinard 1991; Dorai et al. 1998).

The expression "migration system", on the other hand, is used to designate a more encompassing whole, structured by the dynamics of several migration fields, involving feedback loops and path dependency phenomena2. Gildas Simon thus

2. Policy decisions are influenced by a set of past decisions.

defines the migration system as "the articulation and global architecture of migration fields operating on a vast geographical scale" (Simon 2002). This definition is similar to the one proposed a few years earlier by the sociologist Massey et al. (1993) in a landmark article on the main theories of migration. We can then speak of Latin American, West African, European and North American migration systems.

2.3.2. *The region "from below": a circulatory territory*

In recent years, geographers have largely abandoned the notion of a migration system. Instead, they have placed more emphasis on the migratory and transnational circulation initiated by migrants, and on the way in which these have reinforced regionalization dynamics, through the structuring of a "territory of mobility" (Faret 2003), a "multisited territory" (Cortes and Pesche 2013) or a "circulatory territory", to use Alain Tarrius's expression (1992). These expressions reflect the fact that the study of migration is not limited to the settlement of migrants in this or that country, but looks at the multiple forms of relationships (economic, familial, emotional, etc.) woven between the different places that make up migratory spaces.

Laurent Faret's research has highlighted the contribution of Mexican transnational communities to the structuring of a circulatory space between Mexico and the United States. Similarly, Chadia Arab's work on circulation between Morocco and the European Union has identified the existence of a regional circulatory territory specific to migrants belonging to the Aït Ayad group (Arab 2009). Other studies have insisted on the constitution of a Mediterranean space of intense commercial circulation, animated by migrant collectives (Tarrius 1992; Péraldi 2001; Schmoll 2004). We are thus witnessing phenomena of migratory regionalization *from below*, where migration is only one aspect of a broader phenomenon of mobility and transformation of spaces by transnational or translocal communities.

Regionalization *from below* can also be observed in intra-European mobility. Thomas Pfirsch recently studied the role of kinship in the structuring of Italian communities dispersed in Europe. He has shown that Italian transnational families were structured in real households covering several European states. Intra-family relations were structured in particular by care mobilities on an intra-European scale. He studied, in particular, the mobility of grandparents, coming to the aid of their children – now parents – dispersed in several European countries, the "flying grandparents" whose mobility is facilitated by free movement and the existence of low-cost flights. These mobilities can be short, on demand ("whether it's for the flu or because I can't go to school, I'm coming", said one respondent) or long and ritualized, in the form of family tours of Europe lasting several months:

> This is the case of Alessandra, a 74-year-old Neapolitan pensioner. She spends more than 6 months a year away from home, visiting her relatives in Europe. In the spring of 2016, she went for a month to her son's home in Paris, then stayed for a month with her brother in Cologne, who migrated to Germany in the 1960s, and finished the tour with a week's vacation in Amsterdam, staying in a hotel. As a widow, Alessandra did not travel alone, but with friends of the same age, who were also "flying grandmothers", combining family support and tourist trips. Each time, she is hosted by her relatives (she has her own space in her son's Parisian apartment), and takes charge of the daily management (school shuttles, cooking etc.) (Pfirsch 2021).

2.4. The political dimension: a crucial aspect for understanding the structuring of migrant regions

2.4.1. *From the migration system to the migration regime, a strong political connotation*

We have referred to works that describe migrant regions as spaces of circulation and dense and dynamic interrelationships. Other approaches focus more on the political, institutional and governance dimensions of these regional formations.

This is the case with the OECD's definition of a "migration system", which differs in part from the definition proposed by geographers in the previous section. According to the OECD, a migration system is "a set of regulations relating to migration and migrant status that apply across a state or region" (Simon 2002). Applying such a definition, what characterizes the "migration system of the Gulf countries", for example, is not only the particularly high rate of foreign population[3], but also a strong regulatory policy element, that of *Kafala* – a legal system under which every foreign worker must be sponsored, with the sponsor (*kafil*) guaranteeing their migration and temporary presence.

Such a definition is close to what is called the "migration regime". In its reports on migration, the OECD seems to use the terms "system" and "regime" almost equivalently. But the notion of migration regime is itself polysemous (Mützelburg 2019). Some develop a Foucauldian approach, attentive to the dimension of control and coercion inherent in many migration systems (see in this respect the journal

3. The Gulf countries are characterized by extremely high foreign population to total population ratios: 34% of the population or 10.8 million immigrants in Saudi Arabia, 87% of the population or 8 million immigrants in the United Arab Emirates in 2019 (Pison 2019).

Movements – Journal for Critical Migration and Border Regime Studies). In such a sense, the idea of a "Fortress Europe" can be read as the declination, in critical geography, of the notion of a European migration regime or system (Mützelburg 2019).

2.4.2. *The double European migration regime*

As Ettore Recchi (2021) puts it, the European Union is a unique case, since it is based on a "double migration regime". On the one hand, a migration regime closed to the arrival of regular mobility from the global South, through which countries organize themselves to build a common policy base to "protect" their external borders from such arrivals. On the other hand, the institution of a free movement area, since the Schengen Convention, and its enlargement with the entry of the countries of Central and Eastern Europe into the European Union, has been a fundamental variable in the strengthening of regional movement. These intraregional migrations are of different types: more or less qualified, from the South to the North and from the East to the West, they involve workers, students, retirees (Trenz and Triandafyllidou 2017). The increase in recent years of intra-European migration has, moreover, provoked negative attitudes on the side of some European states and contributed, in Great Britain, to the development of an opinion in favor of Brexit (Dennison and Geddes 2018). Yet, it should be remembered that intra-European migration, even after the opening to the East, remains a relatively limited phenomenon: only a minority of Europeans regularly crossing borders are migrants. International migration remains, in the end, a relatively minor mobility phenomenon, despite free movement, or perhaps even by virtue of it: indeed, rather than settling in another country, Europeans favor other forms of mobility to other countries, such as tourism, cross-border movement and short study mobility (less than one year).

2.5. Conclusion: the power of proximity

Region, field, system, migratory regime: the different concepts used to describe the regionalization of migration have both a descriptive and explanatory dimension. They highlight the relational and circulatory dimension of migratory phenomena, the structuring role of communities and social groups in their formation, the importance of proximity in the distribution and concentration of migrants. At the same time, these notions bring to light the constraints and/or facilitating elements of a more political and structural nature, which are at the basis of the constitution of these regional spaces.

Francophone geographers have played an important role in the reflection on the regionalization of migration. They have been able to describe, with statistical data and qualitative research, the regional formations resulting from multiple migration and mobility dynamics, in an approach that is perhaps more empirical than that of English-speaking geographers. Their work has helped to qualify the grand narrative of globalization, which, when applied to the interpretation of migration dynamics, continues to underestimate the place of intraregional migration.

These geographers have often worked in an interdisciplinary manner, in an epistemological context of development and institutionalization of migration studies, which has seen them collaborate with socio-anthropologists, historians, political scientists and demographers since the 1980s. This interdisciplinary approach explains the attention of geographers to the social (migration networks), historical (sometimes post-colonial) and political (fall of the Wall and European integration, borderization) dimensions of the regional configurations observed.

The research presented in this chapter thus points to a complex multifactorial approach to the formation of migratory regions, far from the classic explanations in terms of differentials, or even hydraulic metaphors that are regularly put forward to explain migration. While these simplistic or metaphorical visions may be appealing at first glance, they ultimately contribute to mythmaking and fuel migration anxiety about a spillover from the global South to the North, while ignoring the role and power of proximity in structuring migration dynamics.

Today, with the strengthening of borders and the hardening of migration policies, the geography of migration is undergoing a critical renewal, around approaches attentive to the phenomena of "frontierization" that reinforce, or sometimes counteract, the dynamics of regionalization (Migreurop 2017; Amilhat-Szary and Hamez 2020), opening up new avenues and new sites in terms of the engagement of geographers in the public sphere.

2.6. References

Amilhat-Szary, A.L. and Hamez, G. (eds) (2020). *Frontières*. Armand Colin, Paris.

Arab, C. (2009). *Les Aït Ayad : la circulation migratoire des Marocains entre la France, l'Espagne et l'Italie*. Presses Universitaires de Rennes.

Bahoken, F. (2021). #CarteBlanche sur la migration syrienne. *Carnets néocartographiques*, 7 February 2021 [Online]. Available at: https://neocarto.hypotheses.org/12389.

Cerrutti, M. (2020). 5 salient facts about intra-regional migration in South America. *Migration Data Portal*, 5 March 2020 [Online]. Available at: https://www.migrationdataportal.org/blog/5-salient-facts-about-intra-regional-migration-south-america.

Cortes, G. and Pesche, D. (2013). Territoire multisitué. *L'Espace geographique*, 42(4), 289–292.

De Haas, H., Castles, S., Miller, M.J. (2014). *The Age of Migration*, 5th edition. Palgrave Macmillan, London.

Dennison, J. and Geddes, A. (2018). Brexit and the perils of "Europeanised" migration. *Journal of European Public Policy*, 25(8), 1137–1153.

Dorai, K., Hily, M., Ma Mung, E. (1998) Bilan des travaux sur la circulation migratoire. Direction de la population et des migrations, Paris.

Dumont, G.F. (2009). Le système migratoire méditerranéen. *Outre-terre*, 23(3), 257–272.

Faret, L. (2003). *Les territoires de la mobilité : migration et communautés transnationales entre le Mexique et les États-Unis*. CNRS éditions, Paris.

Flahaux, M.L. and De Haas, H. (2016). African migration: Trends, patterns, drivers. *Comparative Migration Studies*, 4(1). doi: 10.1186/s40878-015-0015-6.

Héran, F. (2018). Comment se fabrique un oracle. La prophétie de la ruée africaine sur l'Europe. *La Vie des idées*, 18 September 2018 [Online]. Available at: https://laviedesidees.fr/migrations-afrique-prejuge-stephen_smith-oracle.html.

International Organization for Migration (2020). World Migration Report 2020. IOM.

Lambert, N. (2015). Une carte à abattre. *Carnets néocartographiques*, 4 November 2015 [Online]. Available at: https://neocarto.hypotheses.org/1963.

Landau, L.B. (2019). A chronotope of containment development: Europe Migrant's crisis and Africa reterritorialization. *Antipode*, 51(1), 169–186.

Massey, D.S., Arango, J., Hugo, G., Kouaouci, A., Pellegrino, A., Taylor, J.E. (1993). Theories of international migration: A review and appraisal. *Population and Development Review*, 19(3), 431–465.

Migreurop (2017). *Atlas critique des migrations en Europe*, 2nd édition. Armand Colin, Paris.

Mützelburg, I. (2019). Régime migratoire. *Hypotheses*, 3 July 2019 [Online]. Available at: https://rm2.hypotheses.org/997.

Péraldi, M. (ed.) (2001). *Cabas et containers. Activités marchandes informelles et réseaux migrants transfrontaliers*. Maisonneuve et Larose, Paris.

Pfirsch, T. (2021). Les tours d'Europe familiaux des grands-parents italiens. In *Dossier "Quo vadis Europa ? La libre circulation européenne à l'épreuve des crises*", Schmoll, C. (ed.). De facto, Aubervilliers [Online]. Available at: https://www.icmigrations.cnrs.fr/2021/04/27/defacto-026–05/.

Pison, G. (2019). Le nombre et la part des immigrés dans la population : comparaisons internationales. *Populations et sociétés*, 563 [Online]. Available at: https://www.ined.fr/fr/publications/editions/population-et-societes/nombre-part-immigres-dans-population-comparaisons-internationales-2019/.

Poinard, M. (1991). Les portugais dans l'emigration : une géographie de l'absence. PhD Thesis, Université Toulouse, Le Mirail.

Recchi, E. (2021). Face aux crises, où en est la libre circulation dans l'Union européenne ? In *"Quo vadis Europa ? La libre circulation européenne à l'épreuve des crises"*, Schmoll, C. (ed.). De facto [Online]. Available at: https://www.icmigrations.cnrs.fr/2021/05/30/defacto-026-02/.

Schmoll, C. (2004). Une place marchande cosmopolite. Dynamiques migratoires et circulations commerciales à Naples. Geography PhD Thesis, Université Paris X Nanterre, Nanterre.

Schmoll, C. (2020). *Les damnées de la mer. Femmes et frontières en Méditerranée.* La Découverte, Paris.

Simon, G. (2002). Penser globalement les migrations. *Revue Projet*, 272(4), 37–45.

Smith, S. (2018). *The Scramble for Europe: Young Africa on Its Way to the Old Continent.* Polity Press, Cambridge.

Tarrius, A. (1992). *Les fourmis d'Europe. Migrants riches, migrants pauvres et nouvelles villes internationales*. L'Harmattan, Paris.

Trenz, H.J. and Triandafyllidou, A. (2017). Complex and dynamic integration processes in Europe: Intra EU mobility and international migration in times of recession. *Journal of Ethnic and Migration Studies*, 43(4), 546–559.

UN DESA (2020). International migration highlights. Report, United Nations [Online]. Available at: https://www.un.org/en/desa/international-migration-2020-highlights.

Valenta, M. (2020). Itinerant labour: Conceptualising circular, serial and stepwise migrations to the Arab Gulf and onwards. *Migration and Development*. doi: 10.1080/21632324.2020.1810897.

Van Houtum, H. and Bueno Lacy, R. (2020). The migration map trap. On the invasion arrows in the cartography of migration. *Mobilities*, 15(2), 196–219. doi: 10.1080/17450101.2019.1676031.

3

Energy Supply: Comparison of Regional Experiences

Angélique PALLE
Institut de Recherche Stratégique de l'École Militaire, Paris, France

3.1. Introduction

Globalization, industrialization and digitalization have made our societies rather dependent on a continuous supply of energy. This supply conditions the lifestyles and production methods of globalized industrial societies, but is questioned in the context of societal debates: regarding the impacts of human activities on the environment, whether it be climate change or global environmental changes (Groupe Cynorrhodon 2020). The dynamics of regional integration that involve the issue of energy supply have two major specificities, which partially distinguish them from other forms of integration. The first relates to the nature of energy flows, which mostly require the construction of production and transport infrastructures whose construction and amortization costs generate significant inertia. The particularly complex technical dimension of interconnection during these integration processes therefore conditions certain political or economic decisions. The second relates to the essential nature of the continuity of energy supply for industrialized societies. This makes energy a quasi-regal domain, in which the delegation of sovereignty or the construction of governance on a regional scale leads to reluctance on the part of nation states.

Between these technical imperatives and the strategic dimension of energy supply, the dynamics of regional integration can be seen as a combined process at several levels or echelons. The technical level, the normative level, the market level,

Regionalization of the World,
coordinated by Pierre BECKOUCHE and Yann RICHARD.

governance and political representations constitute as many layers of energy integrations, which are presented in this chapter. Firstly, the question of the definition of a regional energy integration analyzed through different strata is raised. This theoretical approach, based on the analysis of the European example, is then compared with other regional models, presented through the prism of the transitions underway in the energy sector, which call into question certain choices of the European model. In a context of tensions concerning the security of energy supplies, regional energy integration processes and their governance are studied, between the supposed optimal regional scale and the vulnerabilities it induces.

3.2. Francophone approaches and determining factors of regional energy integration

3.2.1. *Approaches to francophone geography*

Following the example of anglophone geography, francophone geographers have inscribed their analyses of regional integration by emphasizing the socially constructed character of scales (Swyngedouw 2004; Marston et al. 2005). A large number of French geographers approach the phenomena of regional energy integration in their geopolitical dimension. Some of their work concerns the security of supply, considered through infrastructures and discourses. This is the case of Noémie Rebière's work on gas issues in the Mediterranean and on the regional dimension of energy security (Rebiere 2019), or of Sami Ramdani's work on the Visegrad Group's use of a discourse built around a process of sub-European regional integration, for the purpose of controlling gas flows from Russia and the associated infrastructures (Ramdani In press). The construction of integration discourses and their control are also studied in relation to the spaces of margins and "incomplete" integration processes. This is the case of Sophie Hou's work on the Russian–Chinese energy corridors of Eastern Siberia and the Russian Far East (Hou 2019) and Audrey Sérandour on the Andean spaces of the "lithium triangle" (Sérandour 2018, 2020a, 2020b).

Some studies, on the other hand, start from local energy infrastructure to observe its effects on the regional integration process. This is the case of the large hydroelectric dams in China or South America (Sanjuan and Béreau 2001; Mottet and Lasserre 2014; Forget and Velut 2015).

Approaches through the study of norms, legal or technical, are applied to regional energy integration, mainly European. Emmanuelle Santoire considers, for example, the practical, daily and local consequences of a normative change carried out on a macro-regional scale, in a geo-legal perspective (Santoire 2021).

Conversely, the effects on an integrated regional system of decisions taken at lower levels of governance are explored in the work of Angélique Palle, for example, concerning the effects of the German nuclear phase-out on the European electricity system (Palle 2017). Géraud Magrin shifts the focus to actual achievements and questions the lack of integration through the adoption of common standards, particularly in relation to oil resources in the Lake Chad region (Magrin 2014).

Finally, studies on the regional dimension of European energy integration question both its modalities (Palle 2016) and its territorial (Beckouche 2008) or political limits (Nies 2014).

3.2.2. *The determining factors of regional energy integration, a process combining several layers*

In line with this book and with research carried out on European integration, we will consider here the process of regional energy integration from a functional and dynamic perspective, based on infrastructures, common standards, trade, governance processes, energy policies and representations.

From this perspective, regional energy integration is determined by different strata or levels of integration that are more or less dependent on each other. It is difficult to say that there is a linear progression of integration, as Bela Balassa envisaged for trade integration by postulating degrees of regionalism (Balassa 1961). Nevertheless, in the case of trade integration, some of these levels maintain relations of dependence or interdependence. Integration through trade is thus linked to integration through the transport infrastructures that underpin it. There is also a feedback effect. For example, the construction of an electricity or gas interconnection is subject either to an economic profitability imperative to amortize the cost of its construction or to a social or political imperative of access to the resource. Most of these investment decisions are made in the context of regulated monopolies. Once commissioned, these infrastructures tend to be used by the companies in charge of transmission and distribution to the final consumer. They can then be considered as promoting the growth of trade.

While we do not postulate that there is a necessary linear sequence between different layers of integration from a practical point of view (in the sense that we would necessarily precede the other), we can, on the other hand, envisage them as an architecture that arranges elements that range from the most material and the most physically anchored in space (infrastructures) to the most immaterial (representations), and from the most technical to the most political.

The first level of this framework is technical. It concerns the development and connection of production and transport infrastructures, as well as their physical interoperability. This includes, for example, the connection of nonsynchronous areas to each other[1]. This first level is at the base of the integration architecture, because it conditions the exchange of energy between different locations.

The second level is normative. In connection with the technical level, it concerns functional interoperability, i.e. protocols and processes for using infrastructures (rules for connecting new generation units, technical security, etc.). It also includes, in connection with the market level, regulation and common rules for energy exchange (transmission capacity allocations, market auctions, competition rules, etc.). The issue of regulation is also closely linked to the fourth level – political – and conditions certain governance practices. From this point of view, it occupies a key position in the integration architecture.

The third level concerns the market and takes into account economic exchanges. This layer has a commercial dimension, linked, for example, to the opening of markets and the taking of shares in national markets by foreign firms. It also has an organizational dimension with specific integration dynamics such as the coupling of D-1 markets or daily markets. These markets, which ensure the sale of generation and transmission capacity the day before for the next day (D-1) or for the same day (daily), are based on specific auction systems that are harmonized at the time of coupling. This includes, for example, the question of the timing of these auctions.

The fourth level is political. It concerns governance, the legal framework and even the financing of the three previous levels, as well as energy sovereignty – and its possible transfer – in the context of regional integration. At this level, problems of political coordination arise, for example, in the choice of the energy mix. This is the case for the German decision to abandon nuclear energy production. This has had very significant effects on energy flows on the German network, which have affected the stability of neighboring networks to which Germany is largely connected as part of the European energy integration project (Palle 2017). This policy level has close links with the standards and regulation level, especially when it comes to organizing the governance of the technical security of electricity transmission networks.

A fifth level of the integration framework can be postulated, linked to the representations and geopolitical projects or preferences of populations and

1. A synchronous network balances its various components at a single operating frequency. It thus allows AC connections. Two asynchronous networks, operating at different frequencies, can only be connected by direct current lines.

governments. This refers in particular to historical and cultural heritages and to neighborly relations between states, and to the acceptability of certain policies (nuclear, energy transition, etc.). This level has a strong and direct influence on the political level, and therefore an indirect influence on all integration dynamics. Consumer behavior and citizens' priorities can, for example, influence the sizing of the energy transmission network, i.e. an issue that concerns the technical level.

These five levels can constitute the possible criteria for a typology of different regional energy integration models. In proposing such a typology, we are following the constructivist theories of the "new regionalism", which emphasize the development processes of regions rather than their existence and point to the "multidimensional" results of the regional integration process (Hettne and Söderbaum 1998, 1999; Väyrynen 2003).

3.3. Regional energy integration models and the challenge of ecological transition

Using this typology, it is possible to consider and compare several existing models of regional energy integration, which combine the criteria in several ways.

3.3.1. *Comparison of international energy integration dynamics and typology*

The European dynamic, a "bottom-up" technical integration

In the European Union, at the infrastructure level, integration through the interconnection of energy transmission networks is long-standing and very advanced (Lagendijk 2008a). This is evidenced by the electricity grid interconnection objectives currently posted by the EU: 15% of a country's electricity production capacity will have to be exchanged at its borders by 2030. At the normative level, integration is also strong, but not as old or as advanced. There are regulatory agencies and bodies that produce common standards: the Agency for the Cooperation of European Energy Regulators (ACER, created in 2011) and the European Network of Transmission System Operators for Gas and Electricity (ENTSO-E and ENTSO-G, created in 2005 and 2003). But their creation is very recent compared to the length of the integration process. Integration is less advanced at the market level. The integration dynamic is concentrated in certain sectors, such as electricity, but this has little to do with gas and oil. At the political level, the Lisbon Treaty, which enshrines a common energy policy, without prejudging the ability of each member state to determine its national energy mix, attests to the weakness of integration (Bornard 2015). This is the level where integration is weakest, even though it is supposed to play a coordinating role and propose a long-term perspective of the integration process. In terms of representations, the study of Eurobarometers on energy since 1974 shows a growth in the legitimacy of the European level in the field of energy policy among EU citizens (Palle 2016, p. 77). However, the energy-related concerns of European citizens (price, security of supply, sustainability etc.) vary greatly from country to country.

The North American dynamic, relying on politics to ensure security of supply

A priori, the policy implemented in the EU since the late 1990s to integrate the transmission systems of member countries can be compared to what is being done in North America in two respects: first, it is also a vast territory; second, the federated states that make up the United States and the Canadian provinces and territories can be compared to some extent to the EU member states.

Integrating the grids of the various federated states in North America on a macro-regional scale has never been a goal for the federal states and the federated states have very different energy strategies. Some have opted for isolated systems, such as Texas or Florida, as evidenced by the blackouts that affected Texas during the winter of 2020–2021. Others have built interconnection strategies on a regional scale, sometimes across federal state lines, resulting in a diversity of synchronous zones spanning North America. Networks are owned and managed by a plurality of operators who play a much smaller role than the European network operators. The fragmentation of infrastructure and the lack of interconnection between these networks have consequences for trading opportunities. Conversely, the North American regulator (NERC) is much more politically influential than the European regulatory agency (ACER). The management of risks to US supply is the subject of political attention at the federal level via the Electricity Subsector Coordinating Council (ESCC), which has been coordinating a federal response to these risks since the early 2010s, including (unlike in the EU) all US energy sector actors, as well as their Canadian and Mexican counterparts during crisis management exercises (Bossong 2014; Palle 2019).

China's positioning and its regional environment: promoting integration to ensure energy supply, considered in the framework of the Silk Roads project

Faced with these two almost opposite models in the North, it is interesting to compare two possible attitudes towards the regional energy integration dynamics envisaged in the South. The first is observed in China, whose integration policy borrows rather from the European integration model, while being implemented on two scales. On the one hand, it involves building infrastructure to integrate networks within China's vast territory. The government's stated objective is to take advantage of what it presents as territorial complementarities and to use recent Chinese technical developments – in the field of very high voltage direct current lines in particular – to envisage the transmission of energy over very long distances. On the other hand, the particular interest shown by Chinese firms in European network operators, in which they have been investing for 10 years, confirms that the European technical integration model, involving many countries, is a reference for the Chinese authorities. China is thus building a discourse and a project for regional energy integration through infrastructure that is indeed being played out on the scale of a vast regional entity that includes China and neighboring countries. This project is integrated into the broader "Silk Roads" project and can be seen as a political project of regional influence. For example, the energy security of certain neighbors – such as Pakistan – is to be achieved through integration into a regional system in which China would have leadership (Nangia 2019). But this regional integration is also a technical response to the very rapid increase in internal energy demand and the growing political pressure from the Chinese population on the subject of the environment.

An African counter-model? Avoiding integration in the name of an ecological transition

While the Chinese government is promoting an energy integration dynamic, which it presents in particular as a means of satisfying the imperatives of ecological transition in the energy sector, this question is posed in a very different way on the African continent. The implementation of regional integration dynamics on the scale of the African continent has been widely supported since the 1960s by international organizations, which consider them to be a powerful engine of development aid. This is particularly the case for energy and transport infrastructure, the stock of which in the region is the lowest in the world (Jerome and Nabena 2016).

Energy networks are poorly developed and interconnected, and the political dynamics of regional integration in the energy field are weak (Magrin 2014). Rather than promoting regional energy integration, which would require major infrastructure construction, compliance with common standards and a form of shared governance (Akinyemi et al. 2019), a number of works postulate that the decentralization of modes of energy production and consumption would be the most sustainable strategy on the African continent (Lawrence 2020; Zalengera et al. 2020). These works are based on two major ideas. Firstly, the crisis in African governance creates significant challenges for the implementation of a regional energy integration policy that allows access to energy for the population (Gakusi et al. 2015). Secondly, from an ecological point of view, the construction of large production and transport infrastructures, which are intensive in terms of raw materials and require significant development of the territories concerned, could generate more negative environmental externalities than a deliberately local and decentralized system. Finally, these infrastructures, designed on a regional scale, require highly coordinated maintenance on this scale, which is a source of vulnerability. The counter-model, which combines non-integration on a macro-regional scale and a more localized bottom-up development of access to energy, is therefore interesting because it takes the opposite view of the integration of an African "block", still largely thought of as such by international development aid organizations.

Faced with the dynamics of European, American and Chinese integration, the African counterpoint prompts us to question the sustainability of regional energy integration in general.

3.3.2. *Is energy integration a response to the ecological transition?*

Should the energy transition requirement be met by regional integration or by the decentralization of energy systems?

The entire energy sector, on a global scale, is marked by profound changes caused by changes in the geography of producers and consumers, tensions around the supply of energy resources or raw materials (rare earths, lithium, copper, etc.) (Hache et al. 2019), and energy transition policies towards less carbon-based energy sources. In this context of profound changes that are destabilizing the national energy systems of certain states, regional integration processes are sometimes considered as solutions to the tensions in the energy sector. The institutions of the European Union, for example, make integration policy one of the pillars of transition policy in their discourse. They emphasize economies of scale, spatial and temporal complementarity of energy sources according to regions and seasons, and possible solidarity between member states in the face of intermittent renewable energy production.

Other actors are contesting the primacy of regional integration in order to respond to the need for ecological transition. NGOs, citizens' associations and local actors are advocating the decentralization of energy systems and the relocalization of production (Teske 2014). This leads to local opposition to certain renewable energy infrastructure construction projects that are designed within the framework of a European transition policy, with a view to regional integration and centralized production, but which are perceived by local populations as exogenous to their territory and carrying negative environmental externalities (Oiry 2017; Durand et al. 2018).

Governance, the main challenge of regional integration dynamics

In this context, national energy systems, marked by the technical, normative and governance legacies of the 20th century (Beltran 1989; Lagendijk 2008a, 2008b), are put under pressure by a double dynamic of energy integration on a European scale, on the one hand, and a demand for decentralization of energy systems, on the other. In a sector where several actors assert their legitimacy at different scales, the governance of energy systems becomes a subject of questioning.

This is a particularly sensitive issue, because an energy system's mode of governance has direct effects on the technical functioning of this system and on the security of supply. Indeed, as soon as the first level of integration – the technical level of energy production and transport infrastructures – is implemented, there are immediate effects on the higher levels of integration. The interconnection of a gas or electricity network has cascading effects in the event of an incident that can affect the entire system's security of supply. The case of the 2006 blackout in the EU is emblematic in this respect. An incident on a German line affected 15 million consumers in 12 countries of the Union and its immediate vicinity, also threatening the stability of networks around the Mediterranean with which there are interconnections, notably via Spain and Morocco. At the time, the lack of coordination between the network operators, each responsible for their own national territory without a vision of the integrated network, delayed the restoration of the system. Each of these operators had first tried to restore the balance of the system in their own country before coordinating with the regulators of the other countries concerned. Technical integration combined with a lack of political and normative governance thus led to the breakdown of the supply. This power system failure subsequently led to the creation of common standards for incident management, as well as the creation of system monitoring functions on a regional scale (regional security coordinators).

3.4. Conclusion

The study of regional integration in the energy sector is not specifically French-centric. In the foreign literature, especially in English, there is a large body of work on this topic, divided into three themes. The first, classical, is the integration of energy markets, in line with studies on the link between globalization and economic regionalization. The second is the relationship between regional integration and security of supply. The third, more recently, is the possible use of energy integration dynamics in a sustainable transition perspective. French research in geography, while focusing less on the first, is in line with the last two. It has very recently added an attraction for legal issues, explored by young researchers, mainly in a European dimension (Santoire 2021; Ramdani In press).

In the study of regional integration, energy occupies an original place because of the political will expressed (more or less strongly depending on the governments of the territories concerned, but real at all levels) for a transition in the energy sector to meet climate concerns. In this context, the theme of energy integration is increasingly explored through the prism of transition. In the European case, energy integration, initially sought by the EU institutions as part of the construction of the common market, is now, in the discourse of the European institutions, put at the service of an energy transition project that serves the EU's climate objectives (Palle and Richard 2021). Conversely, in the case of Africa, the issue of regional integration, particularly that of infrastructure, is contested in the name of the sustainability of African development (Lawrence 2020; Zalengera et al. 2020).

The regional integration of energy systems is then evaluated from the following perspective: what is the optimal scale for the construction of an energy system that also integrates a sustainability imperative? Answering this question amounts to analyzing integration from criteria of technical, economic or political governance optimality. This is in the vein of economic reasoning on monetary integration and theories on "optimal monetary zones" (Mundell 1961; McKinnon 2000) which have influenced European monetary integration.

The characteristics of energy integrations then present a technical–political particularity. A process of regional energy integration cannot be just one level; it implies an ownership as well as a political project without which the regional energy system created would technically collapse. Thus, Germany's withdrawal from nuclear power, the result of a national political decision in a context of technical, normative and economic integration of European electricity systems (Palle 2017), caused major disruptions among its neighbors, the effects of which were felt politically on a European scale. In the case of regional energy integration, a territorialization process is essential.

3.5. References

Akinyemi, O., Efobi, U., Osabuohien, E., Alege, P. (2019). Regional integration and energy sustainability in Africa: Exploring the challenges and prospects for ECOWAS. *African Development Review*, 31(4), 517–528. doi: 10.1111/1467-8268.12406.

Balassa, B. (1961). *The Theory of Economic Integration*. R.D. Irwin, Homewood.

Beckouche, P. (2008). L'espace économique euroméditerranéen ou la difficulté de passer de l'analyse à l'action. *Strates*, 15. doi: 10.4000/strates.6533.

Beltran, A. (1989). Un réseau électrique national ou régional ? *Bulletins de l'institut d'histoire du temps présent*, 12, 57–63.

Bornard, P. (2015). Politique énergétique européenne : le codicille qui tue ! In *L'énergie en état de choc*, Chevalier, J.-M. and Pastré, O. (eds). Eyrolles, Paris.

Bossong, R. (2014). The European programme for the protection of critical infrastructures – Meta-governing a new security problem? *European Security*, 23(2), 210–226. doi: 10.1080/09662839.2013.856307.

Durand, L., Oiry, A., Palle, A. (2018). La mise en politique de la transition énergétique : la durabilité à l'épreuve des conflits de temporalités. *Temporalités*, 28(28).

Forget, M. and Velut, S. (2015). Grands barrages, frontières et intégration en Amérique du Sud. *Bulletin de l'association de géographes français*, 92(2). doi: 10.4000/bagf.633.

Gakusi, A.-E., Delponte, L., Houetohossou, S.K. (2015). Fostering regional integration in Africa: Lessons from Manantali Energy Project (Mauritania, Mali & Senegal). *Open Journal of Social Sciences*, 3(3). doi: 10.4236/jss.2015.33016.

Groupe Cynorrhodon (2020). *Dictionnaire critique de l'anthropocène*. CNRS Éditions, Paris.

Hache, E., Seck, G.S., Simoen, M., Bonnet, C., Carcanague, S. (2019). Critical raw materials and transportation sector electrification: A detailed bottom-up analysis in world transport. *Applied Energy*, 240, 6–25. doi: 10.1016/j.apenergy.2019.02.057.

Hettne, B. and Söderbaum, F. (1998). The new regionalism approach. *Politeia*, 17(3), 6–21.

Hettne, B. and Söderbaum, F. (1999). Theorizing the rise of regionness. *New Political Economy*, 5(3), 457–473.

Hou, S. (2019). Vers des corridors énergétiques eurasiatiques ? Enjeux d'intégration à l'Est de la Russie. *Bulletin de l'association de géographes français*, 96(3). doi: 10.4000/bagf.5555.

Jerome, A. and Nabena, D. (2016). Infrastructure and regional integration in Africa. In *Region-Building in Africa: Political and Economic Challenges*, Levine, D.H. and Nagar, D. (eds). Palgrave Macmillan, London. doi: 10.1057/9781137586117_6.

Lagendijk, V. (2008a). *Electrifying Europe: The Power of Europe in the Construction of Electricity Networks*. Amsterdam University Press.

Lagendijk, V. (2008b). Histoire de l'idée d'un système européen de l'électricité : projet, progrès, persistance. *Annales historiques de l'électricité*, 6(1), 57–79.

Lawrence, A. (2020). Energy decentralization in South Africa: Why past failure points to future success. *Renewable and Sustainable Energy Reviews*, 120, 109659. doi: 10.1016/j.rser.2019.109659.

Magrin, G. (2014). Autour du lac Tchad, l'intégration régionale dans tous ses états. In *La régionalisation du monde. Construction territoriale et articulation global/local*, Gana, A. (ed.). Karthala, Paris. doi: 10.3917/kart.gana.2014.01.0233.

Marston, S.A., Jones III, J.P., Woodward, K. (2005). Human geography without scale. *Transactions of the Institute of British Geographers*, 30(4), 416–432. doi: 10.1111/j.1475-5661.2005.00180.x.

McKinnon, R. (2000). Mundell, the Euro, and optimum currency areas. Working Paper, Department of Economics, Stanford University.

Mottet, É. and Lasserre, F. (2014). Géopolitique des aménagements hydroélectriques des affluents du Mékong en RDP Lao : développement et intégration régionale. *Canadian Journal of Development Studies/Revue canadienne d'études du développement*, 35(4), 522–538. doi: 10.1080/02255189.2014.966807.

Mundell, R. (1961). A theory of optimum currency areas. *American Economic Review*, 51.

Nangia, R. (2019). Securing Asia's energy future with regional integration. *Energy Policy*, 132, 1262–1273. doi: 10.1016/j.enpol.2019.06.007.

Nies, S. (2014). L'énergie, facteur d'intégration et de désintégration en Europe : bilan du quart de siècle depuis la chute du mur de Berlin. *Herodote*, 155(4), 58–79.

Oiry, A. (2017). La transition énergétique sous tension ? Contestations des énergies marines renouvelables et stratégies d'acceptabilité sur la façade atlantique française. PhD Thesis, Université Paris 1 Panthéon-Sorbonne [Online]. Available at: http://www.theses.fr/s107514.

Palle, A. (2016). L'espace énergétique européen : quelle(s) intégration(s) régionale(s) ? : Réseaux, normes, marchés, politiques, des intégrations à plusieurs échelles ? PhD Thesis, Université Paris 1 Panthéon-Sorbonne [Online]. Available at: http://www.theses.fr/2016PA01H018.

Palle, A. (2017). Les flux électriques européens, de la mise en politique à la politisation ? *Géocarrefour*, 91(91/3). doi: 10.4000/geocarrefour.10229.

Palle, A. (2019). Power networks as targets. Hazards, vulnerabilities and protection of power networks, from the Second World War to the 21st century asymmetric conflicts. *Flux*, 2019/2(116).

Palle, A. and Richard, Y. (2021). Multilevel governance or scalar clashes: Finding the right scale for EU energy policy. *Tijdschrift Voor Economische En Sociale Geografie*. doi: 10.1111/tesg.12481.

Ramdani, S. (In press). Analyse géopolitique du projet Nord Stream 2. Thesis, Université Paris 8 Vincennes, Saint-Denis [Online]. Available at: http://www.theses.fr/s218233.

Rebiere, N. (2019). Géopolique de la sécurité énergétique : la Turquie au cœur des enjeux énergétiques régionaux. PhD Thesis, Université Paris 8 Vincennes, Saint-Denis [Online]. Available at: http://www.theses.fr/s114485.

Sanjuan, T. and Béreau, R. (2001). Le barrage des Trois Gorges. *Herodote*, 102(3), 19–56.

Santoire, E. (2021). Du droit au territoire : enquête sur l'intervention du droit dans les spatialités des systèmes énergétiques (Union européenne, Canada). PhD Thesis, Université de Lyon [Online]. Available at: http://www.theses.fr/s216598.

Sérandour, A. (2018). Quand les savoirs font ressource : constructions sociales et intégrations territoriales. *EchoGéo*, 46. doi: doi.org/10.4000/echogeo.16400.

Sérandour, A. (2020a). Le "triangle du lithium" existe-t-il ? Géographie politique d'une régionalisation andine (Argentine, Bolivie, Chili). *Belgeo. Revue belge de géographie*, 4. doi: doi.org/10.4000/belgeo.43717.

Sérandour, A. (2020b). Le "triangle du lithium" à l'heure globale : marges et intégrations territoriales (Argentine, Bolivie, Chili). PhD Thesis, Université Paris 1 Panthéon-Sorbonne [Online]. Available at: http://www.theses.fr/2020PA01H020.

Swyngedouw, E. (2004). Globalisation or "glocalisation"? Networks, territories and rescaling. *Cambridge Review of International Affairs*, 17(1), 25–48. doi: 10.1080/0955757042000203632.

Teske, S. (2014). powE[R] 2030 – A European grid for 3/4 renewables by 2030. Technical report, Greenpeace, Berlin. doi: 10.13140/RG.2.1.4556.6164.

Väyrynen, R. (2003). Regionalism: Old and new. *International Studies Review*, 5(1), 25–51.

Zalengera, C., To, L.S., Sieff, R., Mohr, A., Eales, A., Cloke, J., Buckland, H., Brown, E., Blanchard, R., Batchelor, S. (2020). Decentralization: The key to accelerating access to distributed energy services in Sub-Saharan Africa? *Journal of Environmental Studies and Sciences*, 10(3), 270–289. doi: 10.1007/s13412-020-00608-7.

4

Transport Systems and Regional Integration

Antoine BEYER
UMR MATRIS, CY Cergy Paris Université, France

4.1. Introduction

As a vehicle for the exchange of goods and the flow of people, the transportation system is a marker of the interdependence of geographic areas. The intensity of these spatial interactions at larger scales then creates coherent groups that are referred to as regional areas, which integrate when these links are closer and more numerous, and when they strengthen more quickly internally than externally. The transport systems that structure these regional areas are defined by the flows, the infrastructures that support them, the economic operators that organize them, and the standards and conventions that shape them, in short, the institutions that frame them. These elements are, in fact, most often concomitant. Because of its massive and hierarchical nature, transport is undoubtedly one of the best ways to illustrate the contours and processes of regional integration.

In this perspective, this chapter shows how francophone geography seeks to think about large regional groupings based on flows of goods and people, and more generally on the control of distance that is the corollary internally and with the outside world. We will try to show how, beyond national legacies and in response to the structuring of globalized flows, a macro-regional scale is taking shape. This can

For a color version of all the figures in this chapter, see www.iste.co.uk/beckouche/regionalization.zip.

Regionalization of the World,
coordinated by Pierre BECKOUCHE and Yann RICHARD.

be seen in the logic of spatial hierarchies around hubs and corridors. The effectiveness of these structures is based on the economic convergence of transport operators and on the definition, imposed or collectively developed, of a common institutional framework.

The close links between history and geography in France have naturally led to the association of preferential exchange relationships that define the great regional spaces over time (section 4.1). Fernand Braudel thus based the construction of civilizational areas on the construction of spatial proximities and discontinuities, in which the geography of transport and trade flows plays a predominant role. Today, the development of a global mobility offer is reflected in the multiplication of flows. A marked hierarchy of transport systems has led to the emergence of large networked regions that form a level of structuring of the world around major air or maritime hubs (section 4.2). At this scale, transport systems are part of systems of increased interdependence between neighboring areas and are changing the logic that has long been the prerogative of states. The need to increase flows on a massive scale is thus leading to the restructuring of regional groupings around transport corridors (section 4.3). The increased coordination of public and private actors appears to be a response to the challenges of the change in scale, both in terms of traffic volumes and coordination of the territories involved. While the impact of these corridors on regional economic growth remains difficult to measure (section 4.4), they do stimulate a dynamic of common regional governance – based on trade and equipment policies, following the example of the European Union within the framework of the TEN-T (Trans-European Transport Network) – which aims to replace the juxtaposition of national networks with a single large European community network (section 4.6). It is therefore interesting to understand how the regionalization of the world has led to the emergence of an institutional form of territorialization, at the intersection of international law, the geography of flows and geopolitics. This process will be illustrated by the presentation of the regionalization of the European airspace and the institutional reorganization that it gives rise to (section 4.7).

4.2. Transport and circulation: key variables in the partition models of the world into large civilizational groups

4.2.1. *A global trend...*

The hypothesis of the control of transport and of a constitutive and regulating distance of large territorial groups is at the heart of Fernand Braudel's work. By placing his analysis in the long term, he established the central concept of a world economy, which he defined precisely as a space integrated by commercial exchanges where trade routes, marketplaces and the system of political control intersect. As the basis for a regionalization of universal history, the world economy thus has an organic spatial structure that is conferred by almost exclusive internal exchanges. For Braudel (1979), the respective extent of world economies is "delimited by the time of movement of its occupants, objects and communications". The extent acquires a coherence ensured by a sufficiently dense transport network to guarantee functional solidarity; the size of regional entities is thus directly linked to the control of mobility systems, from a technical, political or socio-economic point of view. The control of transport and travel speeds defines the maximum extension of sustainable geopolitical entities around a center. Until the advent of mechanized transport, which profoundly increased the distances traveled, a relative stability of travel potentials conferred a geographical durability on human history (Ollivro 2000). The Braudelian space of world economies obeys a concentric organization around an urban pivot, the world city, a commercial heart that organizes the circulation of resources from the margins to the dominated peripheries (Fourquet 1988). The links between world economies have long been tenuous or non-existent. They intensify in the modern era and redraw other regionalizations. This interpretation is partially taken up by the American sociologist Wallerstein (1980, 1984), who makes a more geopolitical reading of it, distinguishing *world economies*, unified by exchanges but diverse in their political structures, from *world empires*, where the imperial type of unity tightly frames internal circulations. The world city is then the capital of the dominant power.

In this all-encompassing perspective of world history articulating time and space, the geographer Grataloup (1996) proposes a dynamic analysis of civilizational areas, in which the question of the mastery of distance through transport remains decisive. He thus considers that the great cultural areas are crossed by two dynamics in tension; on the one hand, the dynamics of diffusion which spontaneously lead a civilization to widen its area of influence; on the other hand, its internal capacity to control distances, i.e. to have the means of communication to overcome the distance and maintain a functional and cultural unity. Organized around a center from which orders leave and towards which surpluses and intelligences converge, these territorial groups must constitute themselves against the tyranny of distance, which is constantly increasing under the effect of the spontaneous extension of a civilizational area. The overcoming of a threshold in the tension between spontaneous diffusion and control of distance leads to the splintering and differentiation of political–cultural sub-entities that end up becoming autonomous (a problem that was highlighted by geographers such as Bowman (1921) and Demangeon (1923)). The maximum extension (we could speak of a *threshold of geographical elasticity*) varies according to topographical conditions (here, limited by the barrier of a high mountain range or desert spaces; there, more extensive, due to the presence of river axes or seas) and the political and military power capable of guaranteeing the security of internal mobilities, but whose sudden weakening distends internal circulations, as illustrated by the eclipses of the steppe empire of the Huns and the Mongols.

Following other research, Grataloup underlines the contrast of the distances that the sea reduces and that land routes stretch. Seas and oceans, far from being barriers to exchange and to the emergence of civilizational groups, have been unifying factors via the reduction of effort necessary to allow the movement of people and goods. Large cultural areas were thus constituted around maritime basins, following the example of the Hanseatic League for the North Sea and the Baltic or the Roman Empire centered on the *Mare Nostrum*. The expansion of Islam around the Indian Ocean is another example, as is the Euro-Atlantic civilization that began to take shape in the 16th century. The construction of regions as civilizational areas is thus linked to the capacities and mastery of a transport whose range was limited for a long time. Technical evolution, stimulated in modern times by new political and commercial ambitions, inaugurates the opening up of closed worlds and their progressive coalescence in a logic of globalization (Grataloup 2015).

4.3. Globalized transport and hierarchical rearrangement of regional units

Successive technological revolutions have changed the geographical organization of the world. The various globalizations largely coincide with major innovations in transportation. Innovations in deep-sea navigation accompanied the Great Discoveries from the 15th century onwards; the mechanization of land transport was concomitant with the Industrial Revolution; the increase of air and sea transport opened up the era of globalization. However, these processes should not be interpreted in a simple causal relationship with the primacy of technology. In many cases, innovation in transport has been driven by political and social demand.

Today, more than ever, maritime traffic represents 80% of the tonnage and 70% of the value of international trade. The world ocean is at the center of a globalized economy and land space. This is the strength of containerization, which reduces delays and transaction costs (Levinson 2016 [originally published 2006]) (Frémont 2011).

The contraction of space induced by a very significant drop in transport costs does not, however, translate into a "flattening of the globe" (Friedman 2005). On the one hand, accessibility gains are not uniform. This leads to differentiations that are potential factors of regionalization. On the other hand, the very principle of structuring a globally integrated transport system is based on powerful concentration effects. At the same time as it makes global flows possible, the transport system relies on a regional base, consisting of highly integrated and hierarchical nodes. This is the principle at work in the network of maritime (Charlier 1996; Ducruet 2016) and air hubs (Cattan 2004; Dobruszkes 2008; Théry 2017): connecting hubs link long-haul flights with short- and medium-haul flights of regional scope, just as major ports enable the transfer between the deep sea and the short sea, or have a gateway function in a continental space. Similar hierarchical approaches are applied to land transport modes (road, rail and river), which are increasingly part of the continuity of maritime and air exchanges, and which serve regional areas. Land transport still accounts for a significant share of total transport costs: between 30 and 40% of the "door-to-door" cost of a container between an Asian manufacturer and its European customer.

The revolution in the organization of transport and its effects on the unprecedented intensification of trade are not contradictory to the process of regionalization. In a way, they contribute to the organization of ever larger continental spaces around a few major port or airport hubs. The scales interlock and highlight the regionalization of the world through globalized transport networks. Within economic areas, logistics specialization is increasing and polarizing storage and distribution functions, combining the proximity of hubs and centrality in relation to consumer markets. The size and contours of logistics areas vary according to the type of product (weight and unit value, frequency and volume of orders, volume of stock, etc.) and the sector considered. The functional reinforcement of massive corridors to serve the hinterland from maritime frontages is the reflection of a process of "port regionalization" (Rodrigue and Notteboom 2005). The effects of coopetition – i.e. the fact that economic actors can be successively in a situation of competition or cooperation – between ports and maritime fronts make it possible to draw highly polarized regional and sub-regional groupings (Chinese maritime frontage around the Pearl River Delta, Shanghai–Ningbo or Bohai Bay; the northern European row dominated by the ports of Antwerp, Rotterdam and Hamburg; the Siamese ports of Los Angeles–Long Beach on the West Coast of the United States). Amplifying this hierarchy of flows, maritime operators are reorganizing on this scale, seeking to ensure global operations in freight and logistics. The same logic explains the structuring of air express freight, where the major operators are developing a system of hubs connected around a main airport per economic region. For example, Fedex organizes its traffic from the hubs of Memphis (America), Paris

CDG (Europe, Africa, Near and Middle East), Guangzhou (China) and Osaka (Asia-Pacific). Finally, for travelers, the main European airports cover preferred destinations. The most important ones (Heathrow or Roissy) provide very open connections with dominant destinations in North America and the Asia-Pacific region, while others offer preferential destinations such as Latin America for Madrid or the Middle East for Istanbul International Airport.

Passenger air traffic presents a mixed picture of regional integration dynamics. On a typical day, the representation of flight density shows the regional polarization of traffic (Figure 4.1). Moreover, a statistical approach allows us to consider the dynamics of each regional grouping. Thus, at the European level, the integration dynamic has not only strengthened since 2004 (Figures 4.2 and 4.3), since internal traffic has grown faster than external flows, but it has also spread to the benefit of secondary airports (Figure 4.4).

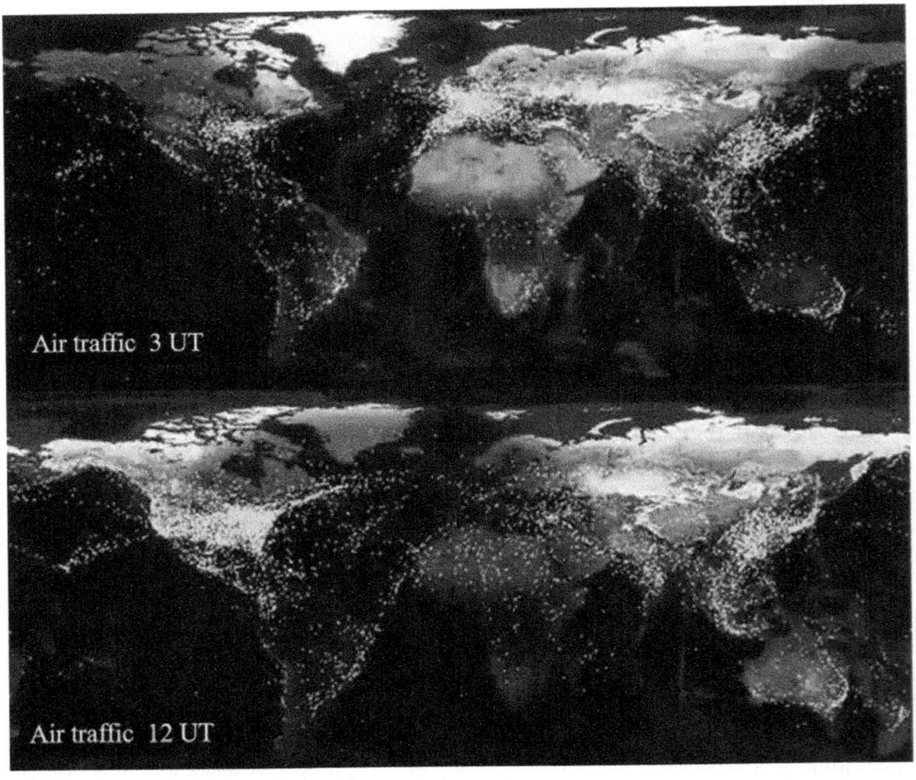

Figure 4.1. *World air traffic (source: Zurich University of Applied Sciences, 2008)*

Evolution of international air traffic in the EU (in 1,000 passengers)

	2004	2014	2019
Outside EU	325,208	502,300	602,361
Inside EU	406,765	603,789	843,296
Inside/outside	1.25	1.20	1.40

Evolution of international port air traffic in the EU (in 1,000 passengers)

	2004	2014	2019
Outside the EU 15	194,139	2,857,586	322,383
Inside the EU 15	191,069	253,044	335,348
Total outside EU	404,453	603,789	843,297
Total inside EU	325,221	462,245	602,366
Inside/outside	0.8	0.77	0.71

EU 15 = the 15 most visited airports in 2019

Table 4.1. *World air traffic (source: Zurich University of Applied Sciences, 2008)*

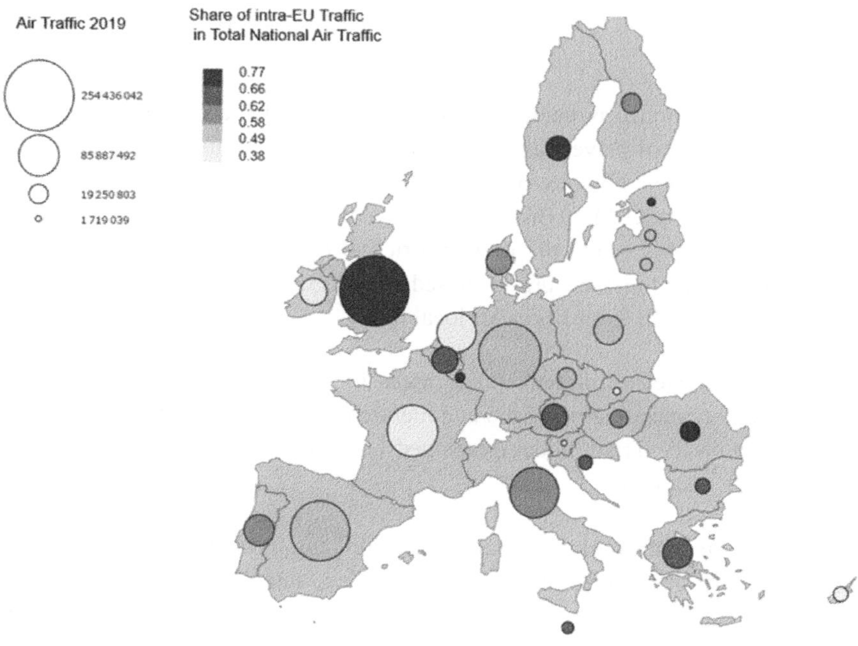

Figure 4.2. *Air traffic in the EU in 2019 (number of passengers, Eurostat data)*

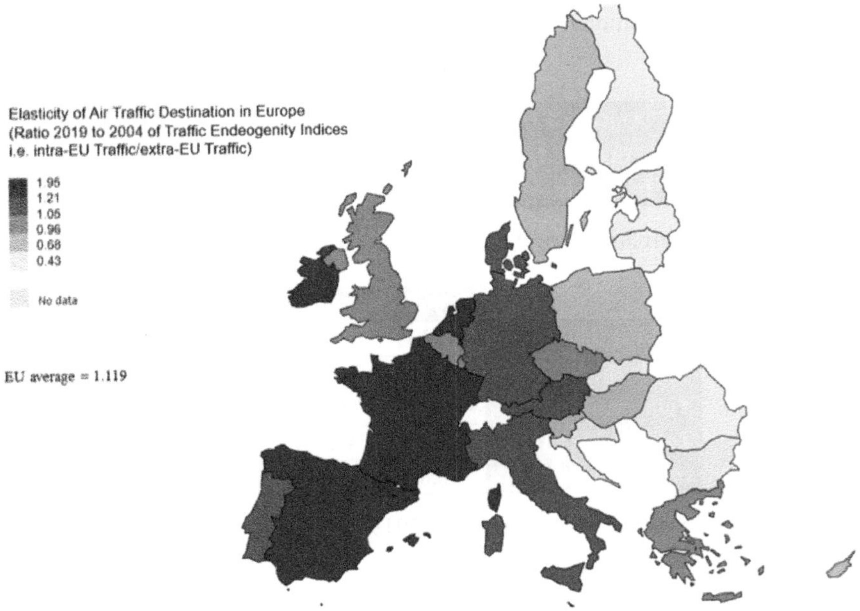

Figure 4.3. *Integration dynamics of air flows in the EU between 2004 and 2019*

Direct capital control is being pushed within each of the economic groupings, reducing the number of players and giving them a regional dimension, which in turn favors an increasingly obvious operational structuring at this scale around large groups (Air-France/KLM, British Airways/Iberia, Lufthansa/Swiss/Austrian Airlines). On a global scale, the more restrictive legislation imposes a system of alliances that brings together airlines based in each of the regional hubs, allowing their partners to benefit from a broader local offer.

What are the relevant organizational areas for global transport operators? The structure of an organization polarized and hierarchized by the major world economic centers, based on the Triad model developed in the mid-1980s by Kenichi Ohmae, remains largely operative.

The trade dynamics of regional integration can be interpreted in light of commercial flows in world trade. There is clear integration if the growth of intraregional flows is faster than outward flows. The WTO's statistics make it possible to account for this over a period of about 15 years. The representation of the evolution of the share of intraregional trade (Figure 4.4) illustrates first of all a strong disparity in regional situations. Western Europe is very self-centered, with more than 70% of its foreign trade directed to its own zone, while, on the contrary, the Middle East, which has traditionally been extroverted by the export of hydrocarbons, does not reach 10% internally. The overall stability of the trade structure does not preclude sometimes rapid changes, as illustrated by the reorientation of the trade of Central and Eastern European countries towards their western partners. At different levels, Asia and Africa have seen their internal trade grow more rapidly than the rest of the world over the past 20 years, while the integration of North America and Western Europe in the early 2000s tending to decline under the impact of Chinese exports.

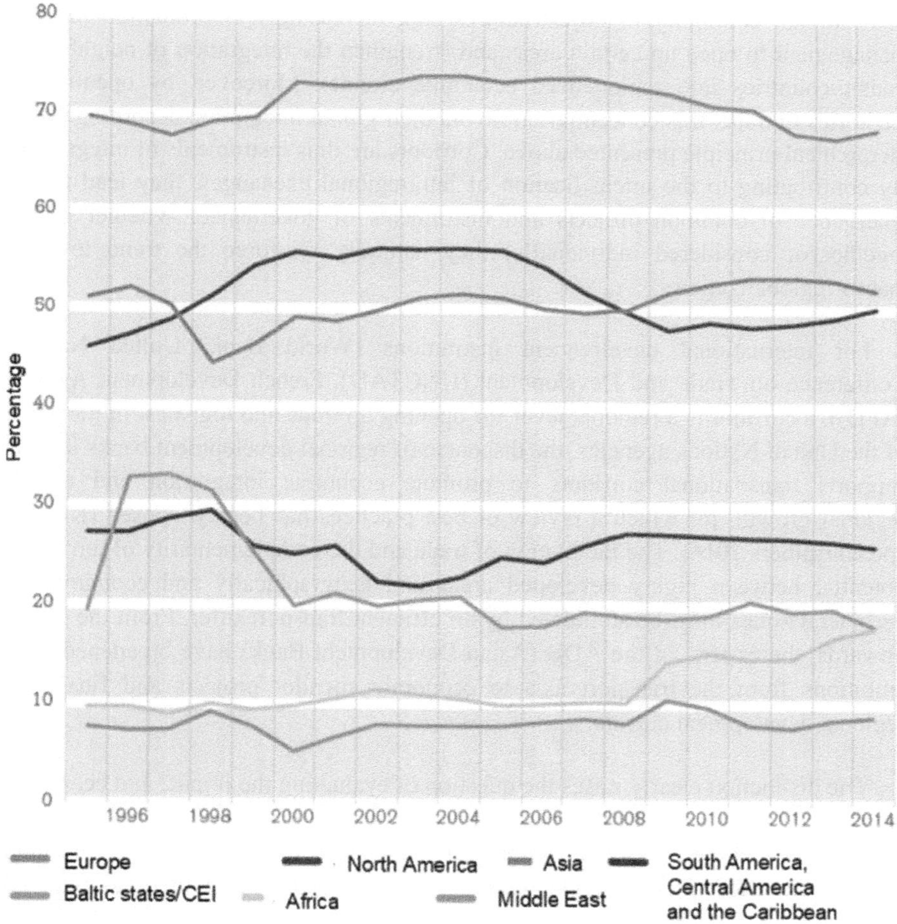

Figure 4.4. *Evolution of the share of intraregional trade in world regions (source: WTO (2015). Statistiques du commerce international, Geneva)*

4.4. Continental transport corridors, at the heart of regional integration projects

The change in the scale of trade is also reflected in the conception of land transport, where the macro-regional scale is now imposed, giving rise to the idea of corridors, i.e. major routes that link several modes of transport, privileged transport corridors, and linear systems of economic interdependence and development. Corridors have an obvious regionalizing power. Applied to freight, they contribute to the efficiency of the transit and exchange system according to the principles of standardization and massification (Alix 2012). This perspective presupposes transnational continuity in infrastructures and, increasingly, coordination in their management to open up certain areas and strengthen the integration of neighboring transit countries into regionalized economic circuits. Moreover, by opening up territories to trade, this concentration favors their global integration according to the hierarchical principle presented above. Corridors are thus instruments of integration. By contributing to the intensification of intraregional exchanges, they lead to the emergence of common projects and instruments of governance. Whether linked together or considered individually, they naturally reinforce the trend towards macro-regionalization.

For international development institutions (World Bank, United Nations Conference on Trade and Development (UNCTAD), French Development Agency (AFD)), a corridor is a regional lever for opening up trade and logistics. In the wake of the United Nations agencies, the discourse of regional development banks largely supports transnational corridors to promote economic integration and ensure regional growth, for which a review of best practices has been proposed (Kunaka and Carruthers 2014). The facilitation of trade and the complementarity of territorial potential between highly developed areas and geographically and economically peripheral zones must be stimulated by an efficient transport offer. From the 2000s onwards, the experts of the ADB (Asian Development Bank) have broadened their ambitions from the transport axis to economic corridor projects and finally to regional development corridors.

The distinction clearly raises the question of evaluating the impact and economic spin-offs of a logistics corridor on the regional integration sought and its structuring effects. Broadening the scope of the reflection leads to questioning the unequal benefits expected. Does the integration it implies lead to growth or to regional development? While the level of investment by states and banking institutions is often high in order to structure the corridors, does it not primarily benefit the major international economic operators? Does the concentration of traffic not run the risk of unbalancing regional groups by draining flows to a few centers, to the detriment of peripheral areas that are made dependent and transit areas that bear the negative externalities?

In West and Central Africa, we observe this projection of heavy investments carried by large regional financial institutions (African Development Bank) to promote transnational transport and trade corridors. In their analyses, Lombard and Steck (2017) highlight the telescoping of the interests of states and large logistics groups that favor small-scale fluidity, to the detriment of the local needs of populations. For landlocked countries, West African transport corridors reproduce an imbalance introduced during colonization by creating an economic dependence on the outside. Massive imports of cheap manufactured goods cannot be sufficiently compensated for by the export of raw materials (Poujol 2017). Integration thus echoes here the construction of a relationship of dependence. For Steck (2015), the role of corridors needs to be reassessed: "What is at stake is the capacity of optimally organized flows to fertilize the territories they cross by being engines of creation of goods and services for all."

Behind the corridor projects on the scale of regional groupings, multiscalar issues are thus emerging, revealing external influences that are amplified by internal tensions. In Africa, as in Southeast Asia, transnational transport corridors are perceived as "instruments of state control over populations living on the margins, either because of their minority status, or because of their illegal activities, or even both" (Fau 2019). Presented as factors of economic cohesion, such corridors play above all a role in potentially destabilizing territories.

Under the guise of economic neutrality, international financial institutions are able to influence regional structuring. Fau (2019) deconstructs the technical approach of developmentalist projects to reveal power strategies. She shows to what extent the shaped operational discourses of international bodies conceal major regional geopolitical issues. In the context of Southeast Asia, there is open competition between Chinese and Japanese economic interests through the financial support given to the construction of this or that corridor. China favors axes that are predominantly meridian in order to open up Yunnan along the Mekong and its western provinces via the corridor with Pakistan. In this, it is opposed to the choice favored by Japan, an east–west and more southern orientation, whose objective is to support the transfer of Japanese industrial activities relocated in Thailand to neighboring countries where the cost of labor is lower. Through the geography of these competing regional corridors, the very geopolitical orientation of the ASEAN is at stake.

In South America, projects for large bi-oceanic corridors were driven by the liberal dynamics of the 1990s. They are supposed to open up inland continental spaces and bring together active coastlines that ignore each other. Programmatic speeches and ambitious maps contrast with the realities on the ground, due to a lack of financial means and the absence of effective economic complementarities (Amilhat-Szary 2003). The isthmic position of Central America has also led to a profusion of projects that seem to impose the region as one of the emerging centers in the global trade network. Yet, their competing projections are rather the mark of disintegration and increased external dependence (Ruckert and Carneiro 2018; Medina 2019).

Elsewhere, on the contrary, the role of transport infrastructure as a materialization of the regional integration drive is more clearly asserted. This is the case for the Gulf Cooperation Council (GCC), a regional organization comprising six monarchies of the Arabian Gulf. To promote internal trade and reduce road congestion, its members approved the construction of the "Gulf Railway" at the organization's 30th summit in late 2009. This mixed freight and passenger project, 2,117 km long, will link the six member states, from Kuwait to Oman, along a coastal route in order to serve an area undergoing rapid urbanization (Cadène and Dumortier 2011). As each state is responsible for the realization of the infrastructure on its territory, a certain wait-and-see attitude has been observed, particularly around the key player Saudi Arabia. Postponed several times due to crises, its complete realization is now expected for 2030.

Considered as internal elements of regional dynamics, the development of transport corridors is just as much the expression of interregional negotiations and projections. The search for balance is expressed through the continuity of land infrastructures or the structuring of maritime services, as much as through the policy of equipping oil and gas pipelines. The financial and banking commitments mobilized are essential levers of this diplomacy of influence. This is well illustrated in the context of the EU's neighborhood policy and its uncertainties (Richard 2016). The major corridors are also fully at the heart of the policies of expanding regional blocs (Cariou 2018). Thus, the unprecedented scale of the New Silk Roads imposes China on two scales: on the one hand, they strengthen its regional power by ensuring its essential supplies and by docking the economies of client states; on the other hand, their extensions are part of a global strategy that affirms the emergence of a new hegemon. Although exceptional in terms of the scale of the funds allocated by China, this practice of investing in corridors can be observed elsewhere, as shown by the role (albeit more discreet) of the EU in the development of initiatives in Africa or in the Integration of Regional Infrastructure in South America (IIRSA).

4.5. The difficulty of measuring the knock-on effects of transport corridors on regional economies

Region, field, system, migration regime: the different concepts addressed for the evaluation of the economic and territorial impacts of transport corridors at the macro-regional level. The concepts bring us back to the classic debate on the expected effects of the positive externalities of infrastructure on the productivity of production factors. The expected multiplier effects are higher for the less well-endowed regions. The structuring effects based on a linear causality of investments on economic growth are widely discussed (Plassard 2003; Bérion et al. 2007; Pumain 2014). This is all the more true for the major regional corridors, which often appear as elements of political or commercial promotion, very partially implemented. The calculation of socio-economic profitability is then based on the yardstick of a specific project and rarely on the scale of the corridor taken as a whole in space and time. Moreover, there are few synthesis studies on a subject dominated by monographs. Nevertheless, reference can be made to the benchmark studies carried out by the World Bank (Roberts et al. 2018) or the UN (2019, 2020; UNECE 2003), in order to identify duplicable lessons, particularly in the least developed countries (Fanou and Wang 2018).

For the gigantic investments linked to the Silk Roads, no macroeconomic assessment of the BRI project for China or for the countries crossed by the land and sea routes seems to have been undertaken to date. While the major ingredients for generating numerous economic externalities seem to be present, the risks of unproductive investment are real. It is clear that what is at stake is not so much the immediate profitability of the infrastructure as the desire to create an area of Chinese political and commercial influence, where the difficulties of financially vulnerable countries can benefit the lending countries.

Even at the European level, and despite the size of the budgets, comprehensive measures of the cost-effectiveness of corridors remain scarce (Schade et al. 2013). For Schade et al. (2019), the completion of the core network in 2030 alone would contribute to a 1.6% growth in GDP from 2017 to 2030 and allow a reduction of 26 million tons of GHG over this period, or 0.67% of the EU27 emissions in 2018, at constant technology. Beyond the average raw data, we must also consider the territorial dynamics that are favored. The lower cost of trade (of which improved transport is only one aspect) makes it possible to take full advantage of the power of economies of scale and the logics of regional productive agglomeration, as highlighted by the New Geographic Economy. These dynamics then lead to the strengthening of the best-endowed regional areas. Conversely, the objective of cohesion calls for a more balanced geographical distribution of growth, a theme that underlies the implementation of regional policies supported by the ERDF. In either case, can the community equipment policy really have tangible and measurable effects? With a European budget of less than 1% of the EU's GDP, of which 3.65% is allocated to TEN-T, we are forced to recognize the anecdotal nature of the financial resources mobilized. This is true even though they are used as leverage to guarantee loans or support national spending. In the interregional economic convergence of the 2000–2008 period and the subsequent divergence phase[1], macroeconomic logic was the determining factor. At the local level, the massive investment in infrastructure made by the EU has nevertheless had significant effects. This has been the case in the CEECs, notably in Poland, but it has reinforced and amplified rather than created the dynamic. For this to happen, the trans-European corridors at higher levels must be linked to local networks in order to have a full knock-on effect.

4.6. TEN-T: political leverage for European integration

Beyond the return on investment which, as we have seen, is difficult to demonstrate, the effectiveness of a regional transport corridor requires the establishment of a governance structure, i.e. a regulatory mechanism that has the capacity to coordinate and converge the efforts of the economic and public actors involved over the long term. In this way, the corridor contributes fully to regional integration through standards and technical–political negotiation. This interpretation presupposes a more institutional approach to the structuring of corridors, as envisaged by Debrie and Comtois (2010). It implies the ability of their managers to define a shared interest in the management of mobility and the handling of goods that contributes to the efficiency of logistics chains and, beyond that, to the establishment of common rules.

The European example of TEN-T (Trans-European Network) is all the more interesting in that the regional institution, the EU, has been provided with legislative means and financial incentives to frame and stimulate the implementation of such corridors. The latter are conceived as instruments that can strengthen territorial cohesion by articulating at a transnational level various common EU sectoral policies: transport policy itself, competition policy, economic development, environmental policy and European spatial planning. Transport corridors are not only the mark of a possible convergence, but also the preferred instrument for the regional institution to act and implement an effective integration policy. At this

1. For data and maps, see Eurostat, gross domestic product (GDP) per capita, purchasing power standard (PPS), by NUTS 2 regions at different dates.

stage, it is therefore useful to recall the context offered by the EU, which is a real player in regionalization, genesis and implementation of this ambition, and the prerogatives attached to it.

In the EU, political discourse closely links the European project and transport supply. Both in the choice of liberalization and in the construction of infrastructure, transport is becoming an emblem of regional integration while contributing materially to its realization. Transport is closely associated with freedom of movement and trade. In particular, the most spectacular engineering structures that cross borders and natural obstacles are called upon, such as the fixed link over the Öresund between Denmark and southern Sweden or the Channel Tunnel linking Great Britain to the continent (Beyer 2011), even though some of these achievements are not directly attributable to European institutions! The latter have only been involved since the Maastricht Treaty (1992) in their financing and since 1993 in the effective liberalization of the sector, with the advent of the Single Market retained by the Single Act of 1986.

Even after the Maastricht Treaty, the definition and implementation of an infrastructure network of European interest has been very gradual, moving from "patchwork to network". Between the two models, we are moving from the articulation of existing national networks to the structuring of major transport corridors, integrated both in terms of the physical offer (continuity and interoperability) and in terms of the institutional unity of their management, which leads to the establishment of common governance frameworks. The idea had been mentioned several times, but never came to fruition (Spaak Report of 1956, *Memorandum Schaus* of 1961, Transport Committee of 1978). The real start was made in 1992 with Article XII of the Maastricht Treaty, which introduced the Trans-European Transport Networks (TEN-T), alongside TEN-E (energy) and TEN-Telecom (information highways), by making community financing possible. The ultimate goal is to create a homogeneous, interoperable network throughout the EU.

The process of reflection will go through various stages until the final plan is established. In 2012, based on ESPON's preparatory work on the interconnection of the main metropolitan areas, the major infrastructures (Core Network) were agreed to be built by 2030 for an estimated amount of €750 billion, and by 2050, the Global Network or Comprehensive Network for €1500 billion (DG Move 2012). In this restricted network, the financial participation of the EU is raised to 20% of the amount invested in current achievements and can be increased to 30% for congested sections, to 40% for cross-border projects and up to 85% of the amounts in countries eligible for cohesion funds. These sums are actually more modest if we take into account the connections and ancillary works that are often necessary.

The regional integration of the transport system is activated by the European institutions at various levels. As we have seen, the first is the financial incentive, while the decision-making power remains in the hands of the states. The Commission also plays a coordinating role. Thus, for each of the nine corridors, it has appointed a coordinator since 2014 to oversee their implementation. In an annual report, the coordinator reports to the Commission on the progress of the projects. Under the aegis of the Commission, infrastructure managers also meet regularly to exchange information and plan work to ensure the continuity of transnational services. Finally, the Commission establishes common technical standards to guarantee the efficiency of the systems and avoid fragmentation of the internal market.

To illustrate our point, we can look at Directive 2014/94/EU, which shows, for example, which levers the Commission can use in the energy transition applied to vehicles. In its desire to have alternative energies adopted, the Commission plans to deploy an infrastructure for alternative fuels throughout the TEN-T core network by 2025. The member states draw up costed national plans, respecting minimum requirements, and inform the Commission of them. The Commission may grant subsidies to facilitate investments that remain the responsibility of the states. Finally, it has given a mandate to the European standardization organizations (ESOs) to ensure interoperability, and it supports consultation between neighboring member states. The core network is indeed the first step in the dissemination of alternative techniques and best practices, as well as the guarantee of regional accessibility.

The management of Eurocorridors reflects the Community's political rationality, which is more horizontal than vertical, oriented towards compromise and knowing how to relay particular interests by putting them at the service of the general interest. Its complex game of influence is part of a multi-level governance. While no one disputes the legitimacy of its action, the institutional and financial resources allocated to it remain modest. It has therefore developed an engineering of management where its influence is increased along the channels it has been able to build at the different levels of decision-making (Beyer 2019). Finally, even before their effective implementation, the merit of the TEN-Ts is the trivialization of the infrastructure planning exercise at the macro-regional level, both in terms of the exchange of information between administrations and national actors and in terms of the ability to get actors who were previously little involved at the international level to work together.

4.7. Towards a regionalization of international transport law?

The construction of a European transport area, as described above, does not stop at the EU's borders. Integration is reflected in the dissemination of community rules and standards to neighboring states. The European airspace will be used here as a guideline to show how structured regional groupings tend to organize larger areas that are recognized by international bodies based on states alone.

Since World War II, the Chicago Convention of 1944 has governed the international air order through the ICAO, a specialized agency of the United Nations. It postulates the principle of complete and exclusive sovereignty of states over airspace and confirms the strictly inter-state nature of international civil aviation law (Grard 2003). In the European context, this principle is mitigated by the inter-state cooperation that developed during the 20th century (Beyer 2020). But it is more particularly the advent of the EU that has been decisive in leading to an integration of airspace at two levels: air traffic control, which, through the "Single European Sky", seeks to overcome territorial fragmentation, and commercial operations with the establishment of an internal market that is completely and freely accessible to resident airlines, under the aegis of the EASA (European Aviation Safety Agency).

This evolution upsets the Bermuda system of free bilateral negotiations between states. The voluntary restriction of state sovereignty mobilizes two legal theories that reaffirm the pre-eminence of community principles (Grard 2003):

– The theory of bottom-up competences states that the principle of undistorted competition is not respected without a unification of the rules of protection of safety, the environment or work; to want the single market is therefore also to extend the homogeneity of the rules in force in other fields, in accordance with the principle of undistorted competition.

– The theory of implicit competences leads the member states to give way to the EU, because leaving the competence to negotiate international air agreements to the member states would substantially affect the common rules already adopted.

In aviation matters, the community rules shake up the legal order inherited from the Montreal agreements. They impose a form of regionalization of the law and, in fact, bring about the emergence of a regionalized airspace that will be placed between the global scale (universality of technical rules) and bi-state agreements (sovereignty of commercial air rights). They render the principle of inter-state commercial agreements between member states obsolete, which is the basis of the ICAO, within the framework of the Bermuda system set up in 1946 by the United States and the United Kingdom, and which served as a model for subsequent agreements. In the Bermuda system, all trade arrangements are exclusively bilateral agreements between a sending country and a receiving country.

This situation is leading the actors in the system to reposition themselves. Member states are called upon to comply with common rules (single market and air law) and adopt a cohesive attitude in the international bodies of the ICAO. The neighboring countries of the EU adhere to the principles and standards of the Community in order to benefit from the technical and commercial advantages of a vast integrated whole: the ECAA (European Common Aviation Area). This form of dissemination of EU law corresponds to a higher level of standardization, which refers to the social and environmental dimension of the European Community regulatory framework. Finally, the other states, and among them the most influential, fear being confronted with unilateral EU regulation. For example, they fought against the 2008 directive (2003/87/EC), which was intended to regulate the greenhouse gas emissions of European and non-European airlines for flights within the EU. This measure was denounced by other countries as abusive in terms of international law. It was finally condemned by the ICAO in 2013. This technical and diplomatic setback for the EU shows the limits of the process of regionalization of air law (Compagnon 2016).

International bodies promote local working groups. The ICAO itself is structured in six regions[2] and collaborates with various regional institutions. But, despite its real influence, the ICAO does not formally recognize the EU as a member, a status reserved for states. The EU policy is then relayed within the international bodies by the member states, which support its orientations with votes and their participation in the working groups. However, on the contrary, there is no systematic opposition from the ICAO to the EU, because the latter adopts the essential measures retained by the international body more quickly and more completely than each state would independently.

The principle of regionalization of air law and the emergence of air sub-areas is therefore a matter of an evolving balance of power. The advances made by the EU are increasingly inspiring other groups of states engaged in economic integration that are relaying experiences of air regionalism: MERCOSUR, the Andean Community or the Caribbean zone within the framework of the ICAO[3] (Correia 2014). In this area, the EU can also be considered as a "normative power".

2. Asia and the Pacific (APAC) in Bangkok, Middle East (MID) in Cairo, West and Central Africa (WACAF) in Dakar, South America (SAM) in Lima, North America, Central America and the Caribbean (NACC) in Mexico City, East and Southern Africa (ESAF) in Nairobi.

3. BAGASOO (Cape Verde, Gambia, Ghana, Liberia, Nigeria), CASSOA (Burundi, Kenya, Rwanda, Tanzania, Uganda), PASO (Pacific), CASSOS (Caribbean), ACSA (Central American Agency of Aeronautical Safety: Belize, Costa-Rica, El Salvador, Guatemala, Honduras, Nicaragua).

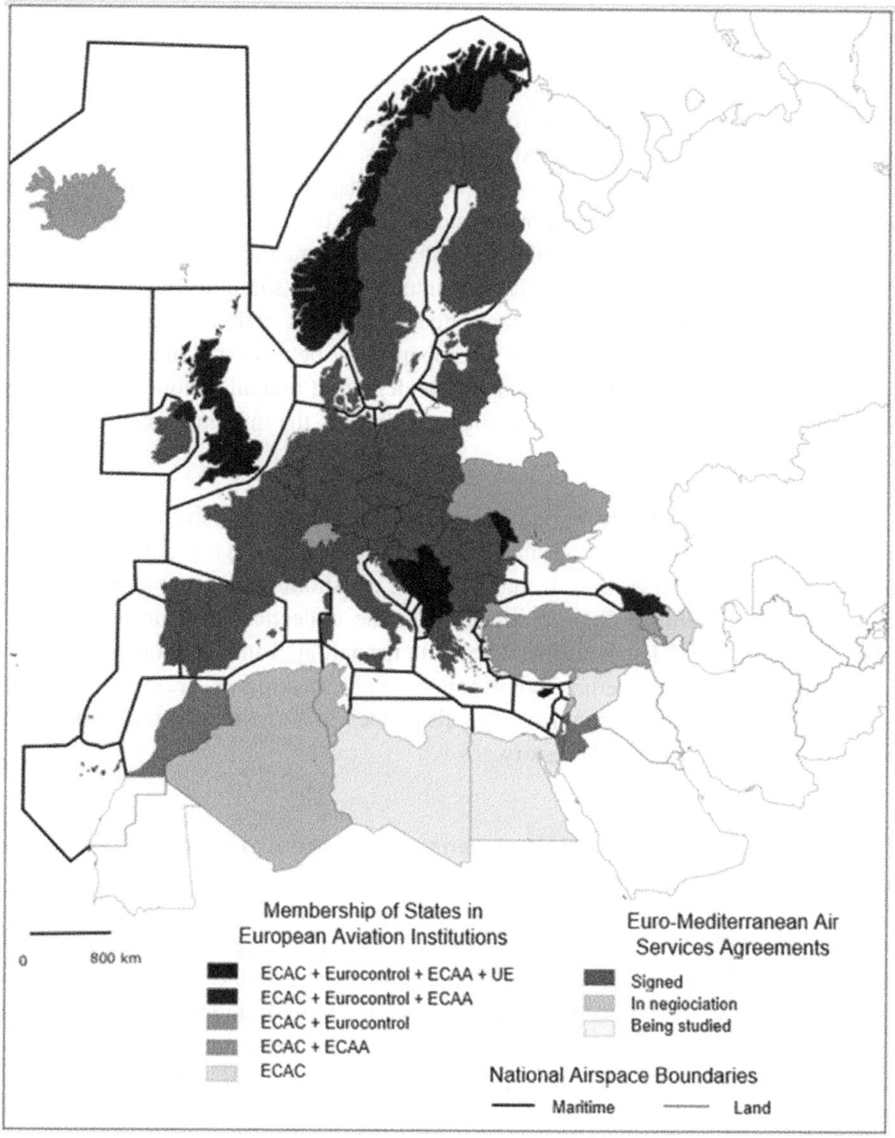

Figure 4.5. *From institutional architecture to regional airspace nesting. Synthesis map (source: Beyer (2020))*

To what extent is the European situation a model for understanding regionalization and its forms?

The negotiation of agreements formalizes the nature and scope of the rights that states grant each other and with external partners. These agreements define a legal and functional space of their own, of which transport systems are the object or indirect indicator. Thus, transport systems can be seen as "hot spots" for understanding how a regional entity is constructed in practice. They offer a grid for analyzing the dynamics and modalities of integration by simultaneously taking into account various dimensions of interrelated exchanges: material flows, physical availability of infrastructure, ease of interoperability between national systems, the existence of regional operating standards, the degree of mutualization of governance (transport policy, infrastructure financing), and finally, the influence and insertion of legal provisions.

The strong and original hypothesis proposed here is that the construction of transport space informs us about the spatial structuring of large regional groupings and their degree of integration. It thus joins the questioning of geographers and can lead to fruitful interdisciplinary exchanges in the dialectical relationship between control of space and legal norms. Their interpretation from a spatial perspective constitutes a very enlightening contribution to the making of regionalization. In return, it assumes that geographical analysis is open to a multidisciplinary approach, both towards law and towards history.

4.8. Conclusion: the multidimensional construction of regional integrations through the lens of transport systems

This chapter shows how francophone geography manages to offer, with regionalization, a new interpretation of transport that is neither sectoral nor solely economic or social. The transport system is used here as a framework for reading the regional integration process. It is expressed at various levels: internal flows that are more intense than external flows, strategic positioning of operators in close interaction, and the eventual implementation of an institutionalization and a regional normative framework. This evolution attests to the capacity to support the regional fact internally, and also to allow its affirmation and dissemination externally.

In the long historical periods before mechanization, the means of transport changed little because of limited and stable resources. They have durably oriented preferential interactions. Thus, accessibility has defined the size of a possible space for exchange control and cultural dissemination. In this respect, from a Braudelian perspective, travel techniques appear over time as vectors for the integration of differentiated civilizational spaces. The intrinsic limitation of transport is thus to be interpreted as a first factor of regionalization. Mechanization increased the capacity and scope of travel. But, far from erasing regional discontinuities, it favors intra- and extra-regional hierarchies.

In the context of the stages of globalization, we are thus witnessing an integration articulated at several levels: around the main hubs, privileged places of interfacing between a region and the outside world and points of convergence within the region itself. This is well illustrated by the contemporary structuring of air and maritime networks. Internally, regional integration is taking shape with the creation of corridors of areas associated with major development policy issues. Although it is often difficult to measure the effects of these corridors on the ground, it is clear that they play a central role in the voluntary construction of political actors and their representations.

The implementation of the European transport network (TEN-T) within the EU framework shows that the reality is already there. Support for the major transport corridors is justified in a regional integration policy. It translates into unprecedented means of intervention on this scale, thanks to institutional levers that provide the project with a normative framework and dedicated financial resources. Beyond these achievements, the EU's capacity for regionalizing action in transport is also reflected in the laws and institutions that govern the transport sector. In international maritime and air transport law, this emergence tends to upset the international order, which recognizes only states as signatories. As a body that produces law and standards, the EU is paving the way to the recognition of regional entities as actors in international law. In fact, the latter is engaged in a process of regionalization.

Thus, in their various dimensions, transport systems are the preferred markers for observing the forms and measuring the intensity of regional dynamics. Of course, regional realities are very diverse. But transport systems in their component parts (flows, actors and institutional logics) provide the means to perceive their contours, estimate their consistency and propose an effective classification on opposable bases.

4.9. References

Alix, Y. (ed.) (2012). *Les corridors de transport*. EMS Editions, Paris.

Amilhat Szary, A.-L. (ed.) (2003). Traverser et utiliser la frontière (Andes/Alpes). *Revue de géographie alpine*, 91(3).

Bérion, P., Joigneaux, G., Langulier, J.-F. (2007). L'évaluation socio-économique des infrastructures de transport : enrichir les approches du développement territorial. *Revue d'économie régionale & urbaine*, 4, 651–676.

Beyer, A. (2019). La gestion du corridor ferroviaire Rhin-Alpes, une mise en pratique de la construction européenne par les transports. *Géoconfluences* [Online]. Available at: http://geoconfluences.ens-lyon.fr/informations-scientifiques/dossiers-regionaux/territoires-europeens-regions-etats-union/rte-t/corridor-ferroviaire-rhin-alpes/@@download_pdf?id=corridor-ferroviaire-rhin-alpes&uid=1b01449677d14edc860e97b2160c0c05.

Beyer, A. (2020). Les institutions internationales, révélatrices des dynamiques régionales. Pour une lecture géographique de l'espace aérien européen et de son intégration. *Belgeo*, 4.

Beyer, A. and Gimenez-Capdevilla R. (2011). La diffraction des interfaces techniques et douanières dans les réseaux ferroviaires ouest-européens (Royaume-Uni, France). *Border Region in Transition XI*, Geneva and Grenoble.

Braudel, F. (1979). *Civilisation matérielle, économie et capitalisme : XVe-XVIIIe siècle*, Tome III. Armand Colin, Paris.

Cadène, P. and Dumortier, B. (2011). *Atlas des Pays du Golfe*. Presses de l'Université Paris-Sorbonne.

Capo-Canelles, V. (2018). Retards du contrôle aérien : la France décroche en Europe. Information report 568, Commission des finances du Sénat, Paris.

Cariou, A. (2018). Les corridors centrasiatiques des nouvelles routes de la soie : un nouveau destin continental pour la Chine. *L'Espace géographique*, 47(1), 19–34.

Cattan, N. (2004). Le monde au prisme des réseaux aériens. *Flux*, 58(4), 32–43.

Charlier, J. (1996). The Benelux seaport system. *Tijdschrift voor economische en sociale geografie*, 87, 310–321.

Compagnon, D. (2016). Expliquer la "crise" entre l'UE et l'OACI sur la réduction des émissions de gaz à effet de serre dans l'aviation internationale. *Études internationales*, 47(2–3), 263–283.

Correia, V. (2014). *L'Union Européenne et le droit international de l'aviation civile*. Bruylant, Brussels.

Debrie, J. and Comtois, C. (2010). Une relecture du concept de corridors de transport : illustration comparée Europe/Amérique du Nord. *Les Cahiers Scientifiques du Transport*, 58, 127–144.

DG Move (2011). Handbook on the regulation concerning a European rail network for competitive freight. Regulation EC 913/2010, Brussels.

Dobruszkes, F. (2008). *Libéralisation et desserte des territoires : le cas du transport aérien européen*. Peter Lang, Brussels.

Ducruet, C. (2016). La spatialité des réseaux maritimes. Contributions maritimes à l'analyse des réseaux en géographie. HDR, Université Paris 1 Panthéon-Sorbonne.

Fanou, E.-H. and Wang, X. (2018). Assessment of transit transport corridor efficiency of landlocked African countries using data envelopment analysis. *South African Journal of Science*, 114(1/2) [Online]. Available at: http://www.scielo.org.za/pdf/sajs/v114n1-2/15.pdf.

Fau, N. (2019). Comment étudier les corridors de développement ? *EchoGéo*, 49 [Online]. Available at: https://journals.openedition.org/echogeo/18066.

Fourquet, F. (1988). Villes et économies-mondes selon Fernand Braudel. *Les Annales de la Recherche Urbaine*, 38, 13–22.

Frémont, A. (2007). *Le monde en boîtes – Conteneurisation et mondialisation*. Arcueil, Synthèse INRETS 53.

Friedman, T.L. (2005). *The World Is Flat: A Brief History of the Twenty-first Century*. Farrar, Straus and Giroux, New York.

Grard, L. (2003). L'UE et le droit international de l'aviation civile. *Annuaire français de droit international*, 49, 492–515.

Grataloup, C. (1996). *Lieux d'histoire : essai de géohistoire systématique*. Reclus, Montpellier.

Grataloup, C. (2007). *Géohistoire de la mondialisation. Le temps long du monde*. Armand Colin, Paris.

Grataloup, C. (2015). *Géohistoire de la mondialisation. Le temps long du monde*. Armand Colin, Paris.

Kunaka, C. and Carruthers, R., (2014). *Trade and Transport Corridor Management Toolkit*. World Bank, Washington DC.

Levinson, M. (2016). *The Box: How the Shipping Container Made the World Smaller and the Market Bigger*. Princeton University Press.

Lombard, J. (ed.) (2017). *Le monde des transports sénégalais*. IRD Editions, Marseille.

Lombard, J., Ninot, O., Steck, B. (2014). Corridors de transport en Afrique et intégration territoriale en question. In *La régionalisation du monde. Construction territoriale et articulation global/local*, Gana, A. and Richard, Y. (eds). Karthala, Paris.

Medina, L. (2019). Les projets de corridors interocéaniques en Amérique centrale. *EchoGéo*, 49 [Online]. Available at: http://journals.openedition.org/echogeo/17868.

Notteboom, T. and Rodrigue, J.-P. (2005). Port regionalization: Towards a new phase in port development. *Maritime Policy & Management*, 32, 297–313.

Ollivro, J. (2000). *L'homme à toutes les vitesses, de la lenteur homogène à la rapidité différenciée*. PUR, Rennes.

Plassard, F. (2003). *Transport et territoire*. La Documentation française, Paris.

Poujol, G. (2017). Les circuits vivriers du corridor Ouagadougou – Accra : conditions d'un développement inclusif. PhD Thesis, Université Paul Valéry Montpellier 3.

Pumain, D. (2014). Les effets structurants des infrastructures de transport. *L'Espace géographique*, 43(1), 51–67.

Richard, Y. (2016). L'Union européenne : un acteur faible dans un voisinage compliqué. *Revue défense nationale*, 791, 49–57.

Roberts, M., Melecky, M., Bougna, T., Xu, Y.S. (2018). Transport corridors and their wider economic benefits. A critical review of the literature. *Policy research working paper*, 8302, World Bank Group.

Ruckert, A.A. and Carneiro, C.P. (2018). Quelle intégration de l'Amérique du Sud ? Les infrastructures de liaison des régions périphériques et les tendances. *Diploweb.com : la revue géopolitique* [Online]. Available at: https://www.diploweb.com/Quelle-integration-de-l-Amerique-du-Sud-Les-infrastructures-de-liaison-des-regions-peripheriques-et.html.

Schade, W., Senger, F., Rothengatter, W., Meyer-Rühle, O., Sean Brouwer, I. (2013). TEN-T large projects – Investments and costs. Study, European Parliament, DG for Internal Policies, Structural and Cohesion Policies, Brussels.

Schade, W., Fermi, F., Bellodi, L., Martino, A., Maffii, S., Schäfer, S., Welter, S., Hartwig, J., Schade, W., Zani, L., de Stasio C. (2019). The impact of TEN-T completion on growth, jobs and the environment – FINAL REPORT. Report on behalf of the European Commission, Karlsruhe, Milan.

Steck, B. (2015). Introduction à l'Afrique des ports et des corridors : comment formuler l'interaction entre logistique et développement. *Cahiers de géographie du Québec*, 59, 447–467.

Théry, H., (2017). Les ailes de la centralité, réseaux aériens planétaires et mondialisation. *Mappemonde*, 119 [Online]. Available at: https://journals.openedition.org/mappemonde/2044.

Wallerstein, I. (1980). *Le système du monde du XVe siècle à nos jours 1. Capitalisme et économie-monde, 1450–1640*. Flammarion, Paris.

Wallerstein, I. (1984). *Le système du monde du XVe siècle à nos jours 2. Le mercantilisme et la consolidation de l'économie-monde européenne, 1600–1750*. Flammarion, Paris.

5

The Regionalization–Globalization Pair: A Reading of the Evolution of World Trade

Gilles VAN HAMME
Université Libre de Bruxelles, Belgium

5.1. Introduction

Globalization has been the dominant paradigm in the social sciences since the 1990s, used as a key to understanding multiple dynamics observed on a global scale. From an economic point of view, it can be interpreted as the combination of several interrelated processes (Vandermotten et al. 2014): an accelerated growth of trade, relative to the level of world production; an intensification of the division of labor on a global scale, fueled by the globalization of value chains, i.e. the use by transnational companies of the diversity of the world space in order to produce each segment of a product in an optimal way (Gereffi et al. 2005; Hopkins and Wallerstein 1986; Grasland and Van Hamme 2010); an enormous growth in financial flows, feeding a sphere of circulation partly disconnected from the real economic dynamics of production of goods and services. Above all, globalization has been interpreted by the most influential authors (Sassen 1991; Castells 1996; Taylor 2004) as the emergence of a world of networks based on urban hierarchies that can replace inter-state power relations.

This dominant reading of globalization as a world of networks calls into question statocentric visions of the world, as well as those that emphasize the relations of dependence between the center and the periphery. However, in no case and contrary

Regionalization of the World,
coordinated by Pierre BECKOUCHE and Yann RICHARD.

to certain predictions, globalization has meant the erasure of the center versus periphery dependency relationships theorized by Wallerstein (1974; see also Van Hamme and Pion 2012), although it has been accompanied by a tilting of the world's center of gravity from the Atlantic to the Pacific (Dicken 2011) through the rapid development of East Asia, in particular China. As for the more or less explicit predictions of the death of the nation-state, the 2008 crisis, among others, has undoubtedly silenced them, without states regaining the levels of control over the economy that they had developed in the post-war period, far from it.

In other words, readings of globalization have advocated a single (global) scale, through networks whose nodes are global cities (Taylor 2004), neglecting the persistence of multiple dependencies between the center and the periphery (Grasland and Van Hamme 2010) and the metamorphoses of nation-states, which nevertheless remain the central actors of globalization. Moreover, with few exceptions, the regional scale has been neglected, not as such of course (there are entire libraries on the EU or other regional agreements), but as an essential dimension of the dynamics of globalization.

In the 1990s, the debate on regionalization was dominated by neoliberal normativity, based on the assumption that trade growth is necessary and desirable in order to benefit as much as possible from the positive effects of the specialization of economies (Krugman 1991; Frankel et al. 1995). From this perspective, regionalization was initially perceived as a threat (Richard and Zanin 2008), since it promoted trade at intermediate scales that would not allow its full benefit to be derived from the global division of labor. However, the figures quickly proved the regionalization versus globalization alternative to be wrong (Anselin and O'Loughlin 1996; Poon 1997). Indeed, intraregional flows have increased sharply in *parallel* with global (extraregional) flows: between 1986 and 2007, the importance of internal trade, as a share of GDP, increased sharply in the EU (from 27% to 43%), in NAFTA (from 6% to 12%) and in ASEAN (from 10% to 34%) while at the same time interregional flows also increased sharply (from 15% to 21% for the EU, from 9% to 15% for NAFTA and from 45% to 86% in ASEAN)[1]. So much so that the World Bank took note of this reality in its 2009 report (Kochendörfer-Lucius and Pleskovic 2009) to advocate the strengthening of regional areas (Van Hamme et al. 2010) as a necessary step in the development of multilateral trade.

However, it would seem that the 2008 crisis did not only bring down the banks or the states of Southern Europe: it also put an abrupt end to all these debates or, more precisely, marginalized them in the field of social sciences.

1. Own calculations based on IMF data (Van Hamme 2014).

In this context, this contribution modestly aims to establish an inventory of global dynamics through the quantitative analysis of international trade, by exploring trade flows taking into account the regional scale. The thesis defended is that of a concomitant and interrelated development between the dynamics of regionalization and globalization: regional free trade agreements have generally been accompanied by a liberalization of trade with the rest of the world. The aim is therefore to assess the dynamics of regional integration over the medium term (1995–2019) through two main measures, along the lines of what was produced by Eurobroadmap (Van Hamme and Grasland 2011): the dynamics of the geography of bilateral trade relations without taking into account free trade agreements; and the dynamics of trade flows internal and external to free trade agreements.

5.2. A methodology for analyzing regional and global dynamics

The evaluation of regional dynamics raises complex methodological questions, which have been explored in depth in certain international research programs, in particular Eurobroadmap (Van Hamme and Grasland 2011), as well as projects funded under the Espon program (Van Hamme et al. 2012; Beckouche et al. 2014). In a first perspective, the regional division of the world is produced through the analysis of matrices of flows between countries, i.e. without any other a priori than the relevance of the state division (Van Hamme and Grasland 2011). This approach, which is more neutral in appearance, does not easily allow us to evaluate the dynamics at work. Another perspective consists of using politically relevant regional divisions, in particular free trade agreements, to measure changes over time in their integration or their links with the rest of the world (Beckouche et al. 2014).

A first way to assess the importance of regional dynamics is to explore the geographical structure of trade relations at the global level. Using UNCTAD data, we first construct a symmetrical matrix of bilateral trade relations, neglecting the direction of the relations (exports or imports) insofar as we seek to assess the intensity rather than the nature of bilateral economic relations. We then unraveled the matrix of the relative weight of countries in world trade by comparing the share of each country in the trade relations of a given country with the weight of this country in world trade. Indeed, the countries of the world tend to trade more with larger economies. So, all other things being equal, the larger the economy, the more it will trade around the world. Thus logically, all countries in the world trade more with the United States than with the Cocos Islands. In other words, the aim is to evaluate the extent to which a country Y trades more or less with a country X, by eliminating the effect of the mass of country X in world trade.

On the basis of this matrix, we ranked the countries by a bottom-up hierarchical analysis according to the proximity of the geographical profile of their trade relations: countries that trade much more than expected with the same country or countries will be grouped together. This method effectively identifies groups of countries with similar profiles, especially bilateral trade within regional areas. However, it has two drawbacks. On the one hand, some countries may have similar profiles without trading with each other. On the other hand, very large players are generally not ranked with countries that trade a lot with them because a country's value with itself is equivalent to the average, which can produce counterintuitive results. For example, although the American continent constitutes a relatively coherent whole, the United States is not necessarily grouped with the countries that trade most with the United States (Canada, Central America).

Secondly, we wanted to evaluate the dynamics of regionalization through free trade agreements. Such an approach makes it necessary to propose regional breakdowns a priori. We have based our work, on the one hand, on the World Bank's division into large regional groupings and, on the other hand, on the spatial configuration of free trade agreements in the early 2010s (see the map in the Appendix). Despite changes in their geography, we have opted to maintain their spatial configuration over the entire study period, even when these regional groupings have lost their reality, such as the Commonwealth of Independent States. We also base our analysis on the most important agreements in terms of trade volume (ASEAN, EU, MERCOSUR, NAFTA), but we also aim to cover the entire planet by taking into account agreements in more marginal parts of world trade such as Africa (Ecowas), South Asia (SAARC), Central America (CACM), etc. As far as possible, we have discarded agreements that have no effect, such as the Arab Maghreb Union or others. These choices are partly arbitrary, for example, in Africa where multiple cross-agreements can exist, as these regional agreements can be so varied in their nature and their application in time and space.

To assess these regional trade relations, we propose two groups of indicators. In the first group of indicators, trade relations are measured in terms of production level, i.e. the gross domestic product of the zone. This classic measure of openness is evaluated here in terms of large regional areas and/or major free trade agreements, distinguishing between trade within and outside the groups thus defined. This indicator has a significant bias, since it compares trade volumes of finished products measured in trade with value added. When goods cross a border, their entire monetary value is counted in the trade relationship, whereas it is possible that, for example, the exporting country has only done the packaging. With the intensification of globalized value chains, these multiple accounts of international trade have probably increased significantly. In the second group of indicators, we compare internal and external trade over time to a given set. While this indicator correctly indicates the dynamics of "internal integration versus regional extraversion", it makes comparing regional sets difficult because it is highly dependent on geopolitical divisions: the high importance of internal trade in the EU compared to NAFTA is largely related to the difference in the geographical configurations of the two sets (three large states in NAFTA vs. 28 medium and small states in the EU).

5.3. The main lessons learned

5.3.1. *A regional organization of exchanges*

Global trade is organized, among other things, according to the dual logic of commercial power and proximity. When the data are stripped of the mass of each unit (see the methodology above), the logics of proximity draw two or three vast groups (Figure 5.1): a European group that still included most of Sub-Saharan Africa in the 1990s but is now reduced to the European space, the Maghreb, and the former USSR; a US group that tends to integrate with East Asia in an increasing way, building a vast Pacific group; and, conversely, an East Asian group dominated by China and strongly integrated with the US group. Beyond the shifts in economic power relations, the main lesson of these treatments is the persistence of proximity as the primary decisive factor of the organization of world trade. All of the groups are organized according to a logic of proximity, reflecting either the importance of mutual trade or the significant influence of a third country in their trade. This preponderance of proximity is reflected in the geography of the major trading powers (China, EU, etc.), all of which are "constrained" by distance (Figure 5.2). Of course, distance and mass do not explain everything: the links inherited from colonial times or other forms of more or less subordinate historical links, geopolitical determinants and the imperatives of supplying raw materials explain many of the features of the geography drawn by the commercial influence of the world's major economic powers.

Nevertheless, the question remains as to why a regional trade organization persists, even though physical and economic barriers have been greatly reduced, in particular the cost of transport. The reason lies partly in the inertial importance of these proximity relations: historically constructed, they persist through relations based on habits and trust between actors. In addition, free trade agreements, i.e. political agreements largely built on the basis of proximity, are also important.

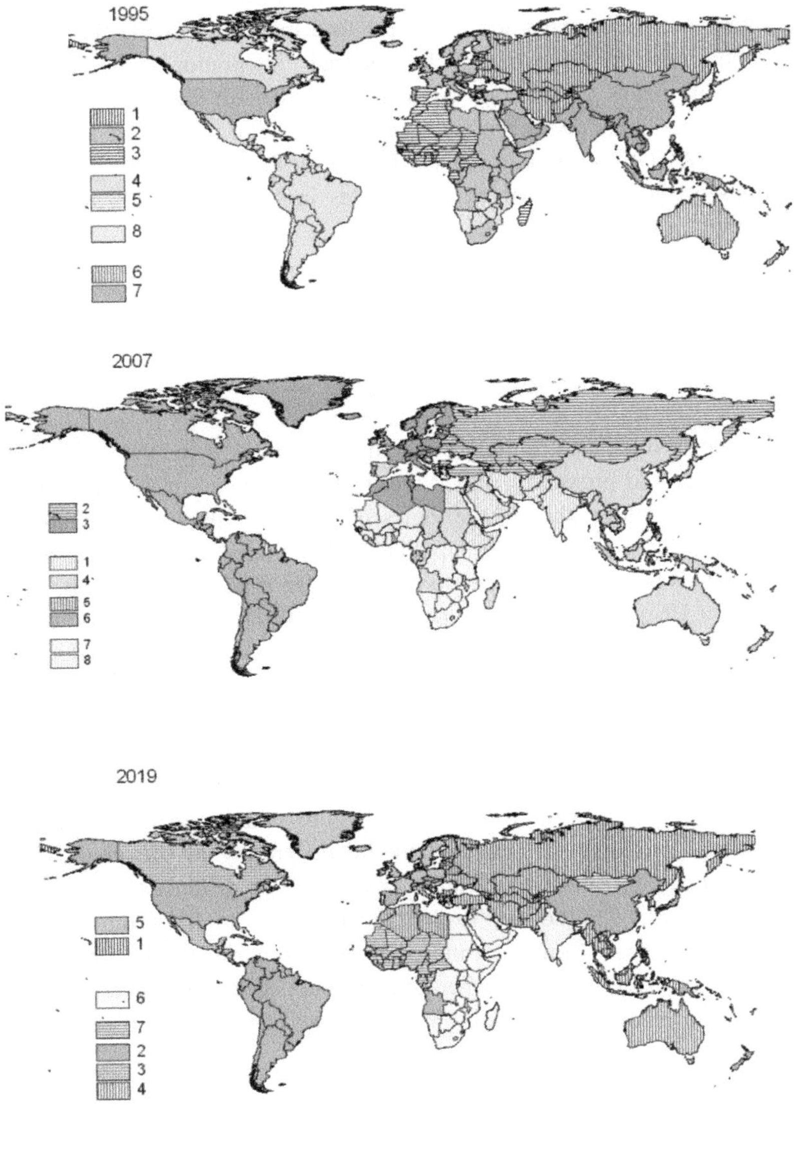

Figure 5.1. *Geographical organization of world trade, 1995–2007–2019 (data source: UNCTAD 2021. Author's processing). For a color version of this figure, see www.iste.co.uk/beckouche/regionalization.zip*

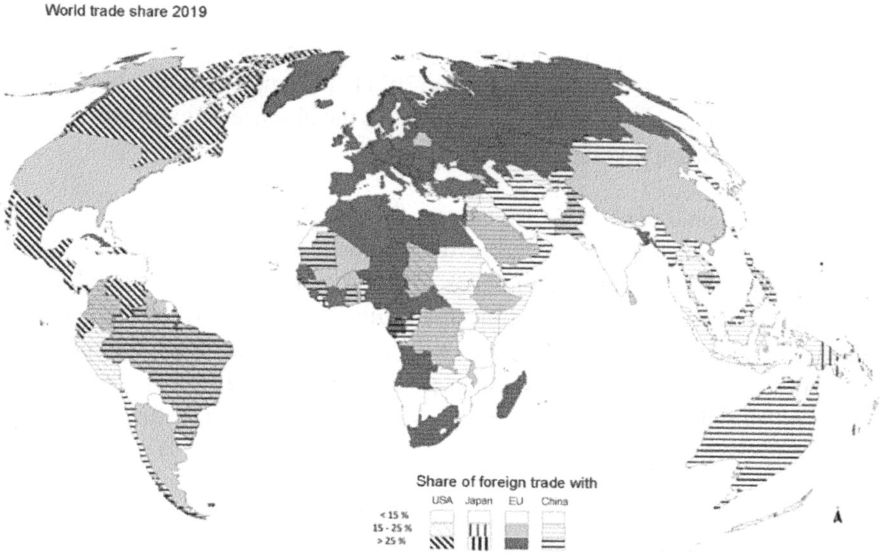

Figure 5.2. *The influence of the world's major trading powers in 2019*

5.3.2. *The deglobalization of the 2010s?*

On a global scale, the 2010s represent the longest period since World War II during which world trade declined relative to the level of production (Figure 5.3). To speak of deglobalization would probably be an exaggeration, given the limited nature of the decline and the very different dynamics observed in different regions. Yet in some parts of the world, the dynamics were enormous. In Asia Pacific, the rate of openness fell from 69% in 2007 to almost 50% in 2019; for China alone, the fall was even more spectacular, from 64% at its peak in 2006 to 34% in 2019, before the Covid-19 crisis. For different reasons, the drop was even more precipitous in the Middle East, Sub-Saharan Africa and South Asia, which returned to the opening levels of the early 2000s. In the most developed countries, however, the trend was less pronounced; and above all, in the EU, there was no sign of deglobalization, since even after 2007, internal integration and trade globalization continued at a high pace.

This deglobalization was not only occurring in the area of trade, but was even more visible when looking at investments, which, after peaks in 2001 and 2007, fell sharply in the 2010s (Abdal and Ferreira 2021). It is not possible, however, to analyze the geography of these flows in detail, in the absence of exhaustive data on the geography of these investments.

Nevertheless, and this is a decisive point, this "deglobalization" was in no way a step backwards, since it only reversed the economic integration produced by the growing integration of world economies to a very limited extent. Therefore, the decline in investment flows did not mean a decline in stocks, just as the decline in trade is limited by the globalization of value chains: the large European and American firms that have organized their production on a regional and global basis for several decades cannot (want to) quickly reorganize their value chains on a national basis, without a significant increase in their production costs.

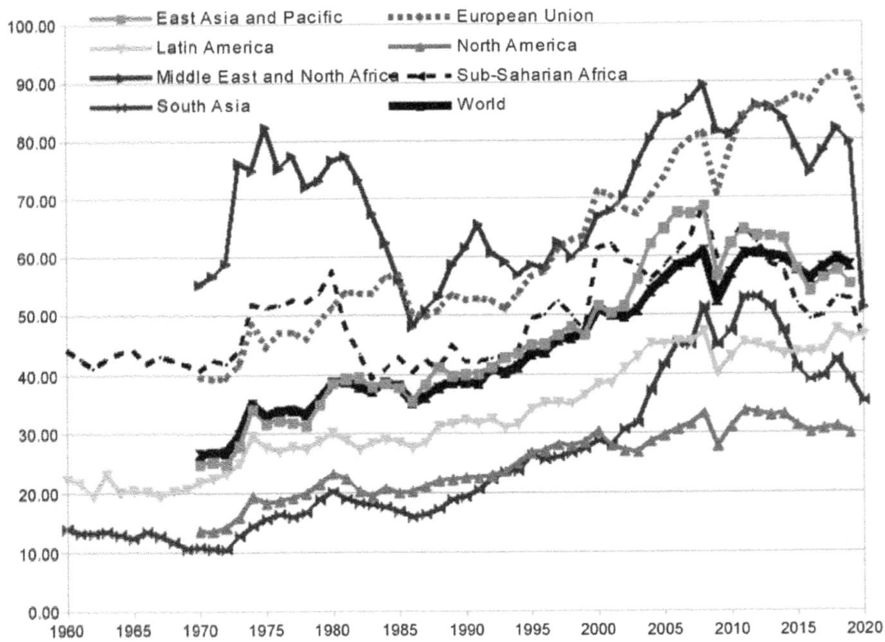

Figure 5.3. *Openness rates (% of total trade/GDP) of the world's major regions, including intra-zone trade, 1960–2020 (data source: World Bank 2021. See: https://data.worldbank.org/). For a color version of this figure, see www.iste.co.uk/beckouche/regionalization.zip*

	Internal trade as a share of GDP (%)					External trade as a share of GDP (%)					Share internal trade vs. total				
	1968	1986	1996	2007	2019	1968	1986	1996	2007	2019	1968	1986	1996	2007	2019
ASEAN	7.7	10.3	22.3	33.5	20.2	24.4	44.6	65.9	85.8	69.7	24.1	18.8	25.2	28.1	22.5
CIS	0.0	0.0	13.0	11.4	14.4	2.0	8.0	29.2	40.8	45.6			30.8	21.9	24.0
EU	11.9	27.3	30.5	42.4	60.0	8.8	14.6	14.1	20.6	29.9	57.5	65.2	68.4	67.3	66.7
GCC	2.9	5.9	5.7	6.0	8.5	64.7	50.6	64.4	76.2	62.5	4.3	10.4	8.1	7.3	12.0
MERCOSUR	1.0	1.4	3.1	4.2	2.9	11.7	13.5	12.4	22.6	23.8	8.0	9.3	20.2	15.8	10.9
NAFTA	2.8	5.5	10.0	11.7	8.2	4.2	9.2	11.7	14.6	13.1	40.4	37.3	45.9	44.5	38.5
Average of all free trade areas	4.4	10.0	14.1	19.7	17.0	7.7	15.3	18.8	28.4	28.5	36.5	39.6	42.9	40.9	37.4

Table 5.1. *Domestic and external trade to the world's major free trade agreements, 1968–2019 (sources: IMF (1968–2007), UNCTAD (1995–2019), World Bank (for GDP)*

The question that arises is twofold: on what scale are trade and the world economy being reorganized? What are the driving forces? With one or two exceptions, regional trade – at least as defined by free trade agreements – has tended to decline (Table 5.1), so that regionalization does not appear to be the level at which the world economy is falling back. Nor is there any reorganization of world trade around the main regional centers: although China's influence is increasing in the Asia-Pacific region, it is barely outpacing the overall growth of its influence in world trade; the same is true for the other major trading powers. This is true in China, as well as in India, and at a more moderate pace in the United States and Russia. Here again, the EU is the exception, with internal trade remaining stable (around 65%) in a general context of strong growth in trade.

Elsewhere in the world, deglobalization may well be more suffered than desired, through the fall in the price of export products such as hydrocarbons, or through the backlash of the refocusing of the world's major economies. Indeed, it is highly unlikely that the "closure" of the African continent will be the product of its own, refocusing on the small national economies that constitute it.

5.3.3. *What are the driving forces of these refocusing processes?*

After 2008, China perceived its dependence on the global market, particularly the US, as a major risk and decided to refocus on its own market by raising wages and increasing household consumption from 35% to 40% of GDP in the 2010s. However, this change was gradual, as it was only possible if there is sufficient technological upgrading of industries; in low-tech, labor-intensive segments, any increase in wages would lead to a decline in competitiveness.

At the same time, in the United States, the debate on reindustrialization was so profound that it was at the heart of Trump's election. His administration took some relatively moderate protectionist measures, extending trends that began at the end of the Obama era and which the Biden administration will not reverse, a sign that there is a deep fracture that goes beyond partisan divisions (Abdal and Ferreira 2021).

Meanwhile, what is happening in Europe? The discourse on reindustrialization has taken hold, but in the absence of industrial policy and/or protectionist measures, it has remained wishful thinking. The globalized strategies of large groups have persisted, and even increased in certain high-level service sectors (Waiengniez 2020).

5.4. Towards a reinterpretation of the regionalization–globalization dynamics

5.4.1. *From the migration regime, a strong political connotation*

Recent developments in the geography of economic flows shed a different light on past developments, i.e. the period of triumphant liberal globalization (1979–2008). This period saw major changes in the world economy, most notably the growing integration of economies on the basis of global value chains.

However, some of the most extreme readings of this period can be definitively denied: the "death of distance", the "death of nation-states" or some readings of regionalization as an autonomous process.

The "death of distance" had already been largely denied by an abundant literature, which had emphasized the importance of agglomeration effects – and even their reinforcement with globalization – or that of proximity as structuring flows on a global scale, despite the reduction in transport and communication costs. The analysis of economic flows has in particular shown the relative cohesion of regional groupings in Europe, America and East Asia, if by cohesion we mean the intensity of exchanges of different kinds based in particular on growing local specialization within enlarged regional markets.

Discourses on the end of the nation-state, which are rarely fully explicit, have been more resistant, especially in a vision of the world based on complex and hierarchical networks, at the apex of which were global cities. Of course, if we look closely, globalization was from the outset a political response to the crisis of the 1970s and was not primarily the result of economic developments (strategies of private actors, technological developments) but rather of political decisions taken by developed states to liberalize economic flows, both regionally and globally: free trade agreements, liberalization of trade within the WTO, end of the Bretton Woods financial regulation and liberalization of financial flows, increasing protection of foreign investments, etc. Once the train of globalization was launched, it became clear that the geographical (global) horizon of private actors was far beyond that of public regulators, giving nation-states an image of impotence. The discourse on the absence of alternatives, i.e. on public powerlessness, is, however, beaten down by reality: states have reacted massively to the financial and economic crisis of 2008; globalization has benefited economies largely driven by states, first and foremost China, whose liberalization and opening up of the economy has always been done in a controlled manner; states are also the only actors capable of dealing with the ecological and health crisis.

From this perspective, regionalism must be read essentially as a process endogenous to globalization. What the World Bank advocated in 2009

– "regionalize to globalize" – is in fact a simple translation of the reality of the 1990s and 2000s: since globalization is encountering certain political obstacles, especially in the periphery, trade liberalization within trade agreements is a stopgap measure. The support of major institutions for regionalization is based on the idea that regional agreement will strengthen competitiveness in peripheral countries by allowing the emergence of larger private actors and specialization within free trade areas. However, this theory is verified in the equations of the New Economic Geography more than in the reality of economic exchanges. Indeed, our data show that most free trade agreements signed in Africa, South Asia or Latin America were hardly accompanied by domestic integration before 2007 (Van Hamme 2014), let alone after. Moreover, regional groupings, and especially the EU, have not only advocated full liberalization of domestic flows (culminating in 1992): the EU also advocates this liberalization within international organizations such as the WTO and imposes it in bilateral agreements with third countries, for which open markets and international investment protection are unwavering cornerstones. The attraction of free access to the world's largest market has convinced many peripheral countries to accept the constraints that accompany the signing of such agreements.

In short, regionalism and globalization have been two sides of the same coin. During the 2010s, when there was a certain degree of deglobalization, regional dynamics slowed down both politically (whereas previous decades had been marked by the establishment of multiple regional agreements) and in the intensity of internal economic exchanges, which were in relative decline (Table 5.1).

The reality is obviously more nuanced, as regional dynamics differ from one region to another. In particular, the dynamics of the EU appear paradoxically – given the EU's leading role in globalization – to be more largely autonomous from the dynamics of globalization. Of course, as we have pointed out, European ideology shares the political readings of the champions of liberal globalization, of which it has been a driving force. But at the same time, it has created an autonomous political entity and, although its elites are steeped in the benefits of liberalism and non-interventionism, the 2010s nevertheless marked the beginning of a shift, for example, in monetary policies and in the need for heavy intervention on the ecological question.

5.5. Conclusion: a look at the production of francophone geography on these issues

As a doubly dominated discipline, a dominated area in the field of social sciences and a francophone discipline facing English-language hegemony, francophone geography has not produced an autonomous conceptualization of

exchanges. For example, the German-based group Allianz announced in 2009 that they were delisting their shares from the *New York Stock Exchange*. Subsequently, from 2010 onwards, the number of IPOs started to rise again, albeit in a very uneven manner across the world (Jurie and Sainteville 2012).

In a study on the shareholder configuration of Asian markets, Ji-Yong Lee observed that the financial crisis caused a reduction in the share of foreign investors, whereas the influence of the latter had increased during the previous decade thanks to the gradual opening of Asian markets after the Asian crisis of 1997. More specifically, the author highlights the decline in the share of non-domestic Asian investors in total capitalization (-38% in value) between 2006 and 2008. Over this period, Hong Kong showed the largest decline in holdings of non-domestic Asian players (-50%), a loss that was partially offset by domestic and US investors. This situation could suggest a decline in regionalization in the evolution of the shareholding structure of Asian markets, mainly to the benefit of domestic investors. However, Ji-Yong Lee (2009) also points out that:

> some countries are taking advantage of the crisis to increase the value of their holdings in the capitalization of neighboring countries: this is the case of Malaysia in Hong Kong (89%), Indonesia (8%), Singapore (22%) and Thailand (17%), Taiwan in Thailand (365%) and Hong Kong (43%), and Japan in Malaysia (138%) and the Philippines (3%).

This observation made at the end of the decade raises questions about the dynamics and intraregional recomposition of capital flows in Asia.

6.3. Regionalization in the organization of the stock exchange activity: the construction of stock exchange groups

6.3.1. *The choice of regionalization for the organization of stock exchange groups?*

At the organizational level, another form of regionalization has been driven by non-state actors, notably through mergers between market firms, i.e. stock exchanges (Figure 6.1). In general, the objective of such mergers is to increase the competitiveness of stock exchange groups in a highly competitive globalized market. The question then arises as to the most appropriate scale, given that the composition of stock exchange groups may have evolved in line with the progress of globalization, the regulatory opportunities offered to them by the various regulators and decision-making bodies, and the economic stakes of the stakeholders. The description of the evolution of stock exchange groups in Europe makes it possible to highlight the logic and dynamics of these mergers, while studying the place of regionalization.

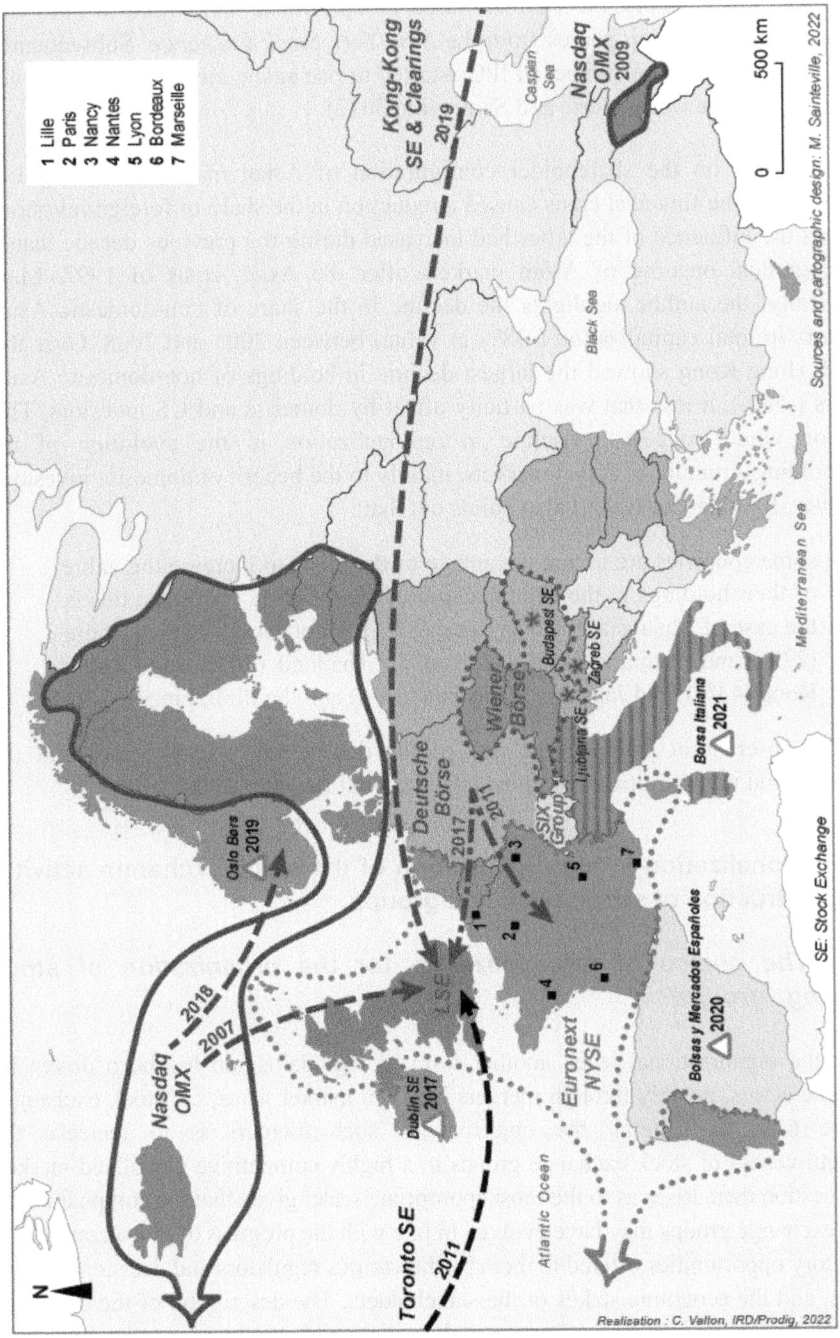
1 Lille
2 Paris
3 Nancy
4 Nantes
5 Lyon
6 Bordeaux
7 Marseille
Kong-Kong SE & Clearings
2019
Caspian Sea
Nasdaq OMX
2009
500 km
0
Black Sea
Mediterranean Sea
Sources and cartographic design: M. Sainteville, 2022
Oslo Børs
2019
Deutsche Börse
2017
2011
Wiener Börse
Budapest SE
Zagreb SE
Ljubljana SE
SIX Group
Borsa Italiana
2021
Nasdaq OMX
2018
2007
LSE
Dublin SE
2017
Euronext NYSE
Bolsas y Mercados Españoles
2020
SE: Stock Exchange
Atlantic Ocean
Toronto SE
2011
N
Realisation : C. Vallon, IRD/Prodig, 2022

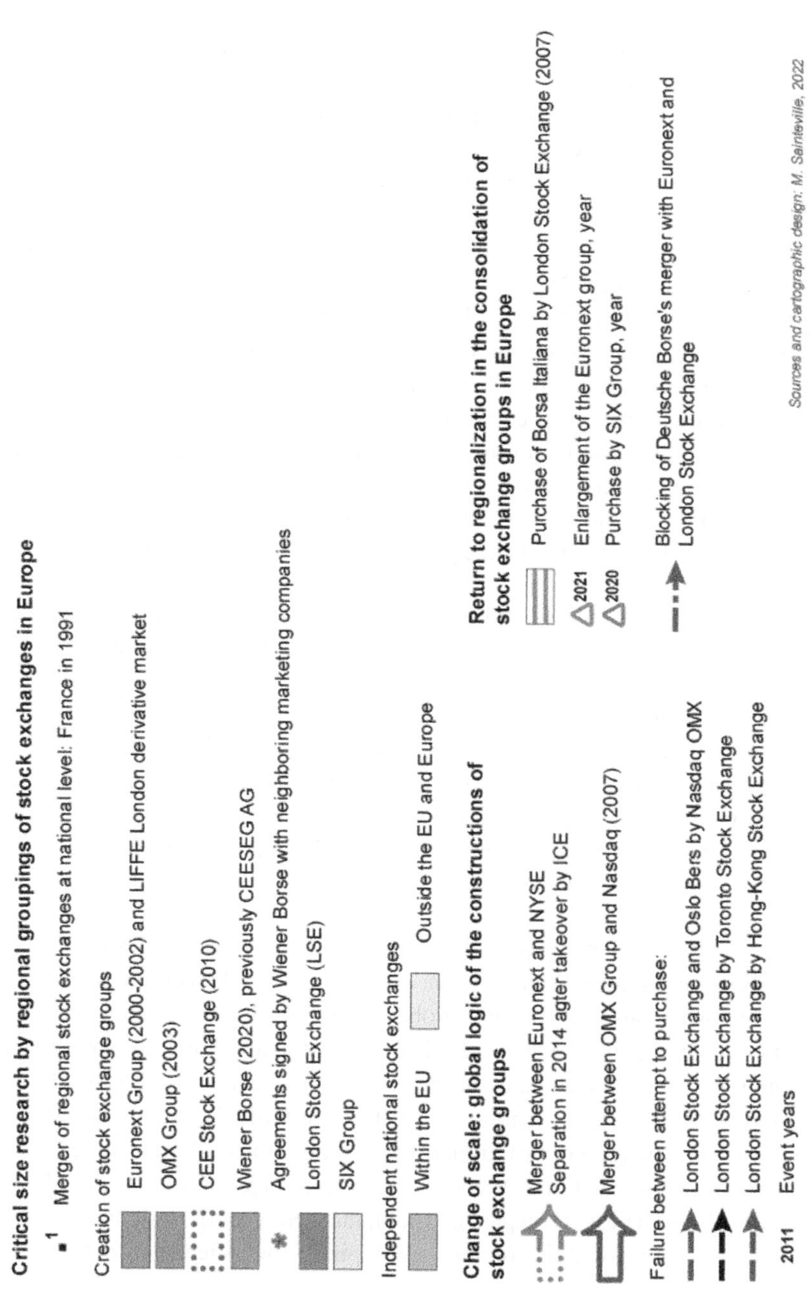

Figure 6.1. *Searching for critical size through regional clusters of exchanges in Europe. For a color version of this figure, see www.iste.co.uk/beckouche/regionalization.zip*

Thus, the consolidation of stock exchanges was developed from the 1990s onwards. The market companies were then seeking to reach the critical size to truly propel themselves into the globalization of the stock market. In France, the regional stock exchanges (Bordeaux, Lille, Lyon, Marseille, Nancy and Nantes) merged with the Paris stock exchange in 1991 and were reduced to the role of a commercial antenna. Beyond the economies of scale brought about by this merger, the purpose of listing shares on the Paris list alone was to prepare for the opening of borders following the signing (1992) and subsequent ratification of the Maastricht Treaty, which founded the European Union. In other words, it was a question of anticipating competition from other European financial markets and financial centers in the context of regional integration. Thus, this reform and the regrouping of stock exchanges on a national scale could be influenced and accelerated by institutional regionalism and the construction of the European Union. Although Switzerland is not a member of the European Union, it embarked on the path of consolidation with the merger in 1996 of the stock exchanges of Basel, Geneva and Zurich, which led to the creation of the Swiss Exchange group (SIX Group since 2008).

During the 2000s, these groupings changed scale with the emergence of pan-European champions in the stock market field, notably Euronext (Paris, Amsterdam, Brussels, Lisbon and London derivative exchanges), followed by OMX Group in 2003, the Northern European market that brings together a group of Scandinavian and Baltic stock exchanges (Stockholm, Helsinki, Copenhagen, Reykjavik, Tallinn, Riga and Vilnius). These marriages have been encouraged by the accession of EU member states to the single currency, the euro. The result is a strategy of integration on a (macro)regional scale and a progressive division of Europe into stock market blocks.

This was followed by a period in which the movement of stock exchange concentration accelerated, and a form of competitive dialectic was established between regional consolidation and global alliances; these two strategies are not necessarily incompatible over time. The first strategy aims to strengthen the positioning on a regional scale, as in the case of the Central and Eastern European stock exchanges with the formation in 2010 of the CEESEG (Central and Eastern Europe Stock Exchange Group) around the Wiener Börse. At that time, the stock exchange group included the Vienna Stock Exchange (Austria) and the stock exchanges of Budapest (Hungary), Ljubljana (Slovenia) and Prague (Czech Republic). In view of this regional grouping, the Vienna Stock Exchange had previously signed a series of bilateral agreements with the neighboring exchanges – notably in terms of sharing technology and financial information, while at the same time setting up equity investments in their capital (Budapest Stock Exchange in Hungary, Ljubljana Stock Exchange in Slovenia). Moreover, the alliance between London SE and Borsa Italiana in 2007 could be perceived geographically as a form of resistance to the offensive of non-European companies on the European stock market. At least, this was the discourse and positioning of Clara Furse, the director of the London Stock Exchange at the time of the merger with the Italian exchange.

The second strategy goes beyond the regional framework. The transatlantic mergers of NYSE-Euronext in 2007 and the Nasdaq-OMX Group, orchestrated by stock market companies, show that global integration was able to short-circuit regional consolidation for a time, despite the fact that it was supported by European political recommendations. In this respect, the geo-economic machinery is seizing up to the extent that the financial interests of the market companies prevail over the pan-European political conscience. The fact that the Lachmann report has not been taken into account is proof of this. In this report, which dealt with the evaluation of stock market consolidation scenarios concerning Euronext, Henri Lachmann encouraged a regional merger with Deutsche Börse over a transatlantic merger with the NYSE (New York Stock Exchange) (Lachmann 2006). In 2007, however, the merger between the pan-European group and the NYSE was completed and gave birth to NYSE-Euronext, which, in addition to the equity marketplaces, also included the London Stock Exchange's derivatives market, LIFFE. At the same time, Euronext's mainframe moved to Basildon in the suburbs of London in order to be closer to its biggest clients, both intermediaries and investors.

This configuration of consolidation on a global scale has given rise to real stock market battles, as illustrated by the competition between the *Nasdaq-OMX Group*, the Dubai Stock Exchange DIFX and the Qatar Investment Authority in 2007. Since then, the Nasdaq-OMX Group has been consolidated (2007), including the stock exchanges of Stockholm (Sweden), Helsinki (Finland), Copenhagen (Denmark), Reykjavik (Iceland) and the three Baltic markets (Estonia, Latvia, Lithuania). After having been rejected twice for its acquisition offer on the London Stock Exchange (LSE), the Nasdaq-OMX Group took over the Armenian Stock Exchange in 2009. Since then, there have been numerous takeover attempts or other mergers that go beyond the European regional framework: the failed merger between the London Stock Exchange and its Canadian counterpart (Toronto Stock Exchange) in 2011, and the failed proposal to buy the Oslo Stock Exchange by the Nasdaq-OMX Group in 2018, after the Norwegian regulator refused. This merger would have enabled the formation of a very coherent Nordic regional sub-group. In addition to these examples, the attempt by Hong Kong Stock Exchange and Clearings to take over the London Stock Exchange (LSE Group) has failed. After the transatlantic experiments, this attempt now brings the Asian exchanges into the race for consolidation on a global scale.

However, the regional configuration does not seem to be an undesirable alternative, and the situation in Europe even seems to support this logic favoring neighborhood and proximity effects. Thus, in 2014, the Intercontinental Exchange group ICE – which took over NYSE-Euronext – split off from the pan-European equity market Euronext, while retaining the activities of London's LIFFE. This short-lived transatlantic merger experiment has given Euronext back its autonomy and can be interpreted as a sign of the refocusing on the European scale. At the same time, it confirms the permanence and relevance of regional integration in the stock market industry. This regional dynamic is reinforced by recent consolidation movements: the Dublin Stock Exchange joined the pan-European group Euronext in 2017, followed by Oslo Børs in 2019 (after the refusal of the merger with Nasdaq-OMX) and the Italian Stock Exchange in 2021, which was acquired from the London Stock Exchange. LSEG wanted to sell the Italian Stock Exchange to meet the requirements of the competition authorities. This sale was won by Euronext, but the competition of Deutsche Börse and SIX Group underlined the economic stake of the merger in a regional context. As part of this acquisition of the Borsa Italiana group and following the momentum created by the Brexit, Euronext moved its main data center (and associated co-location services) from Basildon in the UK to Bergamo in Italy (Aruba's data center) in June 2022. This technology migration then strengthens the regional structure of the exchange group. In the meantime, the Spanish stock exchanges have joined the Swiss SIX Group. These last two examples involve players from non-EU countries (Switzerland and Norway), which underlines that the European stock exchange issue is not limited to the European Union.

While these European mergers seem to attest to regionalization, they remain subject to the goodwill of stakeholders and the oversight of public authorities. In this case, we have already seen the refusal of the Norwegian regulator to integrate the Nasdaq-OMX Group (2018), a refusal to which we should add the refusal of the European Commission for the proposed merger between Euronext and Deutsche Börse in 2011 or for the merger between the German stock exchange and the London Stock Exchange Group in 2017. These two refusals by the European Commission are explained by the fact that these mergers could have undermined the principle of competition on a regional scale.

All in all, the observation of these industrial groupings highlights the coexistence of several organizational and scalar logics, including regionalization. Such a process can also be observed for the *Bourse Régionale des Valeurs mobilières* (BRVM), which groups the markets of eight countries of the Gulf of Guinea and West Africa from its headquarters in Abidjan (Côte d'Ivoire): Benin, Burkina Faso, Côte d'Ivoire, Guinea-Bissau, Mali, Niger, Senegal and Togo. This regional exchange was created in 1996 within the framework of the West African Monetary Union. More recently, the Colombian, Chilean and Peruvian stock exchanges have revived a decade-old merger project. At the end of 2021, they approved a merger with the creation of a regional holding company based in Chile. This merger would enable them to become the second largest stock exchange in Latin America. This new stock exchange group could become effective after approval by the regulator of each country.

6.3.2. *Regionalization favored by regional integration: the case of stock market activity within the European Union*

The institutional dimension appears to be another modality of the regional integration of financial activity. Within the framework of Economic and Monetary Union, provided for by the Maastricht Treaty in 1992, the European Union has equipped itself with regulatory instruments and tools to control financial activities. Thus, the adoption of the euro as the single currency within the euro zone and the implementation of a common monetary policy under the responsibility of the European Central Bank since 1999 have favored financial exchanges. Similarly, in Africa, the West African Economic and Monetary Union (WAEMU), created in 1994, has provided a favorable framework for the growth of capital flows and the creation of the *Bourse Régionale des Valeurs mobilière*s (BRVM) in 1996.

In 1999, regulatory and institutional tools were also adopted to implement the principle of free movement of capital within the single market. Thus, European players, whoever they may be, can carry out financial operations in all EU Member states under identical conditions, in accordance with the Financial Services Action Plan adopted by the Commission in 1999. Among the measures of this FSAP are the harmonization of accounting standards, the construction of a common legal framework for securities and other financial instruments, or the abolition of barriers to the raising of capital within the Union. More recently, the harmonization of the functioning of financial markets within the European Union has been the subject of successive directives: MiFID (Markets in Financial Instruments Directive), and then MiFID II, which should strengthen investor protection by better controlling the organization and conduct of players on these markets.

If these factual, organizational and institutional aspects contribute to reinforcing the regionalization of financial activity practices, we can question the relevance of the regional scale with regard to the structural and technological evolution of stock market practices.

6.4. What role for regionalization in the evolution of stock market activity?

In the context of this questioning of the regionalization of stock market practice, the effect of technological and institutional developments should not be underestimated. The new challenges linked to the speed of information transmission and the new discriminating role of distance in certain sectors of market finance (HFT) are pushing certain actors to move closer together and tighten the practice of exchange in narrower spaces, on a local or regional scale (in the intra-national sense). This raises questions about the future of the remote functioning of exchange within the framework of regionalization as we have considered it up to now, i.e. the organization of activities on an intermediate scale between the nation-states and the world. At this stage, this contribution does not intend to provide a clear-cut and definitive answer. Our objective is rather to study the place of regionalization in the renewal of the stock exchange activity and to draw the attention on the complexity of the articulation of the scales in the spatial organization of the stock exchange activity. Indeed, the challenges remain, at different scales: the interconnection between the major financial centers (notably with submarine fiber optic cables) on a global scale, the consolidation of the networks of players and the possibilities of exchange on a macro-regional scale to reinforce the depth and attractiveness of the markets while ensuring the integration of the financial centers in globalization, and the requirement for speed in the context of HFT, which requires a reduction in the distances between the players of the exchange. Below, the description of this new exchange modality will underline the spatial stakes in order to be able to analyze them in the perspective of regionalization.

During the 2010s, the convergence between computer engineering and financial markets gave rise to a new practice, that of HFT. Trade is evolving and is now essentially carried out by algorithms, programmed to react almost instantaneously to market changes and to send mass orders. The functioning and principles of this new trading modality are discussed by Alexandre Laumonier in his essays *6*, *5* and *4* (Laumonier 2014, 2019). The efficiency of these "robot-traders" relies on their reactivity and on their ability to execute their orders before those of their competitors. In fact, these actors must benefit from reduced latency times, which makes it crucial to have access to reliable and efficient information transmission technologies, in particular through the use of fiber optic networks, microwaves or laser technology (Sainteville 2016). Speed, the ratio of distance traveled to time, has become a real issue for intermediaries (*traders*), as trading platforms execute orders according to their arrival order. This new demand for speed in trading – now timed in microseconds and where every nanosecond counts – has precipitated the evolution of stock market networks and their spatial registration.

A first spatial evolution is the concentration and the bringing together of exchange entities on a local scale. Thus, the computerized order books (the "robot-traders") of financial intermediaries have been installed as close as possible to the electronic trading platforms of traditional exchanges and other "alternative trading platforms", Multilateral Trading Systems (MTS) such as BATS or Direct-Edge, which are now authorized by the Regulation National Market System (RegNMS) in the United States and by the European Markets in Financial Instruments Directive. In practice, these platforms and the intermediaries' algorithms are installed in vast data centers such as those of ICE (previously NYSE) in Mahwah (USA, 40,000 m^2) and Euronext (ex-NYSE-Euronext) in Basildon (Essex, 30,000 m^2) about 50 km from London. It should be noted, however, that following Brexit and the purchase of the Italian Stock Exchange by Euronext in 2021 (Figure 6.1), the data center of the pan-European exchange has been moved to Italy in the second half of 2022, not far from the financial capital of Milan, more precisely to Bergamo. While this move reinforces the regional logic, it also highlights the institutional dimension of the challenge of consolidation within the European Union.

In these data centers, the intermediaries' computers are directly connected to the execution platform by dedicated fiber optic cables, which makes it possible to secure the network and optimize the transmission of information. Thus, we are seeing an increase in the number of data centers that host trading activity, such as Savvis NJ2, which hosts several trading platforms in Weehawken, New Jersey. Hosting companies also offer *proximity hosting* solutions, i.e. the installation of servers in separate buildings, but close and well connected by computer networks to those of the trading platforms. The stakes of time and network control are so high that HFT companies such as Morgan Stanley, Goldman Sachs, Getco and other hedge funds do not hesitate to pay rents of several tens of thousands of dollars per square meter per month.

In addition, HFT and the need to control speed and distance lead to the operation of exchanges on a local scale. Such a concentration of specialized actors can be observed, for example, between the states of New York and New Jersey in the United States (see map Sainteville 2016). Just a short distance from Manhattan's financial district, the proximity and connectivity between exchange platforms allow robot-traders' algorithms to implement strategies specific to high-speed trading, such as "front-running", for example. Without going into detail, front-running is a stock market technique used by an intermediary to make money by taking advantage of price discrepancies in the execution of an order transmitted by one of his clients and of which he has prior knowledge. This type of practice is based on the principle of arbitrage, on the asymmetry of information and on the ability to act at very high speed between the different stock exchange platforms. *Time is money*... In practice, the location of high-frequency (HF) traders in the Weehawken or Secaucus data centers is very advantageous for performing arbitrage, due to their proximity to Manhattan and to their positioning in the middle of multiple trading platforms, as well as between the Intercontinental Exchange (ICE, ex-NYSE) facilities in Mahwah and Nasdaq in Carteret (see Figure 6.1 and Sainteville 2016). Indeed, while the performance of networks and IT solutions is essential for HF traders to assert themselves against their competitors and sweep the exchange, proximity and small distance are undeniable assets (O'Brien 1992; Cairncross 1997, 2001).

In view of all of these facts, it is true that the new requirements of HFT seem to call into question the possibility of long-distance trading in a macro-regional framework unless technical subterfuges are used to manipulate the distance, as proposed by Brad Katsuyama in *Flash Boys: A Wall Street Revolt* (Lewis 2014). For all this, the macro-regional scale, like the process of regionalization, retains its relevance for certain modalities of operation and organization of stock market activity, particularly for forms of exchange that do not necessarily rely on speed. Within the framework of the increasingly complex stock exchange activity, we can see that several complementary organizational and scalar logics remain, among which we count regionalization.

6.5. Conclusion: the power of proximity

From the perspective of the initial questioning of this book, the geographical observation of stock market activity provides some useful insights. Although financial activity has developed networks between actors on a global scale, the regional dimension retains its advantages. In terms of capital raising and securities listing, the majority of issuers are still domestic firms. But the indicator of cross-border listings shows a certain regional preference on the part of foreign issuers, i.e. a relational preference driven by the logic of proximity.

Beyond the practices of exchange, the consolidation of market companies into stock exchange groups is mostly taking the regional route in Europe. Although smaller-scale groupings are still possible, as the transatlantic example of Nasdaq-OMX Group shows, the facilities offered by the fact that the various players (issuers, investors, ISPs, market firms, etc.) belong to the European Union – in particular through the homogenization of regulations – support regional strategies. However, the European challenge of stock market practices and industrial organization is not limited to the European Union alone, as shown by the recent mergers between Euronext and the Oslo Stock Exchange or between SIX Group and the Spanish stock exchanges; Norway and Switzerland are not EU Member States.

All in all, the reflection on the modalities of stock market activity – notably through the observation of HFT – invites us to situate the (macro)regional fact within the framework of a complex activity that plays on the complementarity of scales according to the economic stakes.

6.6. References

Ascher, F. (1995). *Métapolis, ou l'avenir des villes*. Odile Jacob, Paris.

Auvray, T. and Granier, C. (2009). La crise financière en Europe : vers une convergence des modèles de détention actionnariale ? In *Géographies de la finance mondialisée*, Dupuy, C. and Lavigne, S. (eds). La Documentation française, Paris.

Cairncross, F. (1997). *The Death of Distance. How the Communications Revolution Will Change Our Lives*. Harvard Business School Press, Boston.

Cairncross, F. (2001). *The Death of Distance 2.0. How the Communications Revolution Will Change Our Lives*. Texere Publishing Ltd, Knutsford.

Carroué, L. (2015). *La planète financière. Capital, pouvoirs, espace et territoires*. Armand Colin, Paris.

Dupuy, C. and Lavigne, S. (2009). *Géographies de la finance mondialisée*. La Documentation française, Paris.

Géneau de Lamarlière, I. and Sainteville, M. (2008). Une Europe financière entre consolidation régionale et intégration mondiale. In *Atlas de l'Europe dans le monde*, Didelon, C., Grasland, C., Richard, Y. (eds). La Documentation française, Paris.

Jurie, V. and Sainteville, M. (2012). Europe in the arena of the global financial integration. In *The Networked Europolis: Changing Spatial Relations of the European Territory in Globalization*, Pain, K. and Van Hamme, G. (eds). Edward Elgar Publishing, Cheltenham.

Labasse, J. (1974). *L'espace financier*. Armand Colin, Paris.

Lachmann, H. (2006). Evaluation des scenarios de consolidation boursière impliquant Euronext. Report, Paris Europlace.

Laumonier, A. (2014). *6/5*. Zones Sensibles, Brussels.

Laumonier, A. (2019). *4*. Zones Sensibles, Brussels.

Lee, J.-Y. (2009). Le maintien de la diversité du capitalisme en Asie. In *Géographies de la finance mondialisée*, Dupuy, C. and Lavigne, S. (eds). La Documentation française, Paris.

Lewis, M. (2014). *Flash Boys: A Wall Street Revolt*. W.W. Norton & Company, New York.

O'Brien, R. (1992). *Global Financial Integration: The End of Geography*. Pinter Publisher, London.

Sainteville, M. (2009). Structuration, organisation et territorialisation de l'espace économique par l'activité boursière, ses flux et ses réseaux. PhD Thesis, Université Paris 1 – Panthéon-Sorbonne.

Sainteville, M. (2011). Les paradis fiscaux dans la mondialisation boursière – Firms, stock exchanges and geography. The role of tax havens in stock-exchange activity. *L'Espace politique*, 15(3) [Online]. Available at: https://journals.openedition.org/espacepolitique/2180.

Sainteville, M. (2016). Activité boursière : réseaux, lieux et territoires à l'épreuve du temps. *Flux*, 106(4), 27–42.

Sainteville, M. and Géneau de Lamarlière, I. (2008). Une Europe financière entre consolidation régionale et intégration mondiale. In *Atlas de l'Europe dans le monde*, Didelon, C., Grasland, C., Richard, Y. (eds). La Documentation française, Paris.

Savy, M. (1998). TIC et territoires : le paradoxe de la localisation. *Les Cahiers scientifiques du transport*, 33, 129–146.

Terrien, B. (2009). Une nouvelle norme de construction et de diffusion des statistiques d'investissement direct. *Bulletin de la Banque de France*, 177, 61–73.

7

The Runet, a Region of Cyberspace?

Kevin LIMONIER

Institut Français de Géopolitique, Université Paris-8, Saint-Denis, France

7.1. Introduction

In its most literal sense, Runet refers to all networks, digital services and platforms where the Russian[1] language is used to communicate. It should be noted that there is no equivalent term for the other major linguistic segments of cyberspace: if we speak of Runet, we do not speak of Franet to designate, for example, the French-speaking Internet. This is because the Russian-speaking segment of cyberspace has sufficiently remarkable specificities to distinguish it from the rest of the digital networks. The Runet has historically had its own social networks and search engines, which induce a specific user experience quite distinct from the rest of the global Internet, dominated by American or, more recently, Chinese platforms. For a long time, the term Runet has designated a universe characterized by a set of cultural and technical practices that are more or less spatially locatable, since most of its users are located within the borders of the former USSR – with the notable exception of the Russian-speaking diasporas of Europe, Israel and North America. A real digital *topos*, the Runet has taken on a geopolitical significance as the Russian government has sought to control its pipes. At the same time, the use of the Russian language as a marker of identity and memory in many post-Soviet conflicts has given the Runet a particular meaning – to the point of leading the Ukrainian government, for example, to ban the use of

1. This chapter was written before the war in Ukraine, which began in 2022.

Regionalization of the World,
coordinated by Pierre BECKOUCHE and Yann RICHARD.

Russian platforms for reasons of national security, against the backdrop of the annexation of Crimea and the civil war in the Donbass.

From a *topos*, the Runet has become a geopolitical representation in the sense of Lacoste (2014), i.e. an image of space corresponding to a specific political project. Now, this representation is based, to a large extent, on the vision of the post-Soviet regional space as it is told today from Moscow, in order to legitimize a set of strategies of appropriation of the space that Raviot (2017) has summarized under the term "Kremlinocentrism", i.e. "the centripetal dynamic [aimed at curbing] the danger of a breakup of the Russian state and nation and its scattering to the *peripheries*: to the periphery of the world (of the concert of nations), but also the 'overflow' of the 'Center' by the 'peripheries'".

As Raffestin (1980) argued, the strategy of spatial appropriation that underlies Kremlinocentrism requires both the mastery of discourse (of information, in the sense of Deleuze) and that of networks, be they infrastructural, political or economic. The mastery of discourse is based in this case on the celebration, in Russian, of a glorified and idealized common past. With the control by Russian interests of major transport corridors, gas distribution, networks of power or wealth, these narratives constitute the core of a dynamic of regional reintegration (Mareï and Richard 2018) by the center, i.e. by Moscow. This regional reintegration of the post-Soviet space around the interests of the current Russian power is opposed to other dynamics where some former republics seek, with varying degrees of success, to tie themselves to other regional groupings (such as Estonia with the Scandinavian area, for example) and/or to build themselves into nation-states. The result of this opposition between centrifugal and centripetal dynamics is a state of permanent geopolitical conflict in which memory, language and identity constitute objects of rivalry at least as important as other political, economic or geostrategic aspects.

The Runet has become a major component of this Kremlin-centric regional integration and of the conflicts it generates: control of the Internet guarantees both control of the discourses that circulate on it, and control of the power structures that are deployed, born or supported on it. Before going further in the analysis of this phenomenon, it is necessary to specify what is meant by "cyberspace". It refers to the Internet and, beyond it, to all digital data exchange networks (Douzet 2014). The Internet is the network of networks, which it connects to ensure the transport of data packets (pixels, texts, audio data), thanks to routing points. The architecture of the Internet is all the more complex because it does not obey a single scheme that a central tutelage would deploy. To better understand it, we represent it as a stack of layers (Ventre 2011). The information exchanged between users defines the upper layers. Raw data packets exchanged between machines flow through the lower layers. These layers interact, as the former cannot function without the latter. In practical terms, this means that a social network cannot function if its servers are physically destroyed, or if the logical routes that connect it to the rest of the Internet are cut off, as was the case on October 4, 2021, when Facebook, WhatsApp and Instagram "disappeared" from the network for nearly six hours.

All layers of cyberspace are now affected by the regionalization dynamics of Kremlinocentrism, which extend beyond the borders of the Federation and encompass the entire former USSR. Countries such as Kazakhstan, Azerbaijan and Armenia remain almost entirely dependent on Russia for their connection to the outside world. This means that data destined for these countries generally passes through Russian routers or servers, over which state control has increased considerably in recent years. Conversely, this increased control has also accentuated certain dynamics of resistance. In areas such as Ukraine, Georgia or the Baltic States, the segments of cyberspace that are under Moscow's control are seen as potential threats to sovereignty or national security, against the background of geopolitical conflicts with Russia. In both cases, we are witnessing a "digital projection" of the power relations inherent in the dynamics of emancipation that prevail on the margins of the "near abroad", in relation to the regional integration driven from Moscow.

The objective of this chapter is to highlight these dynamics within cyberspace, whether they promote regional integration from the center or limit it – especially from the margins. Given the nature of the subject matter, the chapter can be read as a methodological demonstration. Indeed, despite its largely immaterial character and the impossibility of representing it according to the rules of classical cartography, cyberspace does not escape spatial dynamics, of which it has become one of the main structures in our hyperconnected societies (Douzet 2020).

7.2. From the cultural to the political Runet: appropriation of the "upper layers" of cyberspace

It is at the level of the "upper layers", and more particularly of the platforms of intermediation that the appropriation "by the center" of the post-Soviet digital spaces is the oldest and the most remarkable. By platform of intermediation, we mean all of the online platforms that allow the interaction of the users (Grumbach and Frénot 2014). These include social networks, as well as large search engines or encrypted communication applications such as WhatsApp or Telegram. On a global scale, the market for intermediation platforms is dominated by American players, the famous GAFAs. Even though the dominance of some American digital empires such as Facebook (which owns Instagram and WhatsApp to name but two) tends to decrease slightly in the face of the success of Chinese platforms such as TikTok, their situation remains almost monopolistic everywhere – with the exception of two regions: China (where many western services are blocked) and the Russian-speaking area. In this region of the world, Russian intermediation platforms are still used by a majority of Internet users: whether it is the search engine Yandex, the social network Vkontakte or the messaging platform Mail.ru, these local services are favored by Russian-speaking users.

This is a unique situation in the world, since the domination of these local platforms, unlike China, owes nothing to censorship organized from above: American platforms have never been blocked in Russia (with the notable exception of LinkedIn), even though access to some of them is beginning to be restricted, in a context of authoritarian tightening of power. This is notably the case for Twitter, whose access has been "slowed down" in the country since mid-2021. The reasons for this Russian digital exception are numerous and can be found in the recent history of the region. In the 1990s and 2000s, the post-Soviet zone was of little interest to major western digital investors, who saw it as an insecure, corrupt and unrewarding market. For a long time, GAFAs shied away from Russian-speaking markets. For example, until the early 2010s, Google's algorithm was not calibrated to understand Russian declensions. This made searches on this engine and in this language very approximate, so Russian-speaking Internet users got used to using the Yandex engine, which was programmed for Russian. However, despite the lack of interest from western platforms, the post-Soviet space quickly became a hotbed of the Internet, as early as the mid-1990s. At that time, the former USSR was the only non-western zone to experience a massification of Internet uses. The population had an excellent level of understanding of computer mechanics, and the nascent digital industry provided excellent opportunities for enrichment – more or less legal – in countries in deep crisis since the fall of the USSR. As a result, local platforms quickly developed to meet the needs of Russian-speaking users: founded in 1997, Yandex quickly became the undisputed leader in indexing Russian-speaking content, while Vkontakte, founded in 2006, became the reference social network.

The term "Runet" was coined to describe this unique economic, technical and cultural situation. In the early 2000s, the term referred to the entire Russian-speaking Web and its strong cultural specificities: use of local platforms, as well as intense circulation of pirated or illegal content. Like the post-Soviet societies in crisis, this cultural Runet was the realm of resourcefulness, DIY and the circumvention of all sorts of rules, whether legal, moral or technical. There were entire shopping malls in

Moscow dedicated to the trade of illegal electronic products[2], while powerful cyber mafias flourished there, such as the *Russian Business Network*, which is credited with being the author of most of the credit card hacks that took place in the world in the early 2000s.

Over the course of the 2010s, this cultural Runet and the platforms that embodied it came to the attention of Russian power, which has since indirectly taken control of a majority of them, so that they have become instruments of Kremlinocentrism, its vision of the post-Soviet space and its regional integration strategy. This dynamic of appropriation began around 2012. This date corresponds to the re-election of Vladimir Putin for a third term, which marked a conservative turn based on a Russia that would defend a certain number of "moral values". At the time, tens of thousands of Russians marched in the streets of all the country's major cities against Putin's return to power, when he was Prime Minister and Dmitry Medvedev was the President of the Federation. These demonstrations, the largest the country has seen since the fall of the USSR, were mainly organized via social networks, over which the state and its security services had almost no control. This episode contributed to making the Internet a threat not only to the stability of the regime, but also to the integrity of the "moral values" that the Kremlin defended in a staging of Russia as a "conservative power" threatened by the influence of a west considered decadent. In other words, the network was seen as the privileged vector of western influence capable of perpetrating new color revolutions[3] in Russia or in its "near abroad". In November 2012, Russia suggested reorganizing the governance of the Internet around the model of the United Nations, and therefore of the states sitting in the General Assembly, which would allow the network to be subject to sovereign control. This proposal, which went against the multi-actor governance on which the management of the Internet has been based since its foundation, was rejected by the United States and its allies on the grounds that it could increase the power of control and digital censorship of authoritarian regimes. Russia took advantage of the Snowden affair, which revealed the extent of Internet surveillance carried out by the US National Security Agency (NSA) in June 2013. Not only was Russia able to grant political asylum to Edward Snowden, but it also found reinforcement in its demonstration that the Internet was a major instrument of influence in American hands.

---

2. This was the case for the Gorbushka market, in the west of Moscow. This market, which covered several hectares and was a major venue for the resale of rock music at the end of the USSR, later specialized in the sale of pirated computer products and the development of hacks that were resold on CDs.

3. Popular movements in Eastern Europe, which, often with western support, brought about changes in government at the beginning of the 21st century.

Between the denunciation of the American hegemon, the defense of moral values and the imperatives of national security, the Internet became the object of a triple convergence in Russia, leading to the takeover of the major national platforms by interests close to the government. The platforms that had developed outside the orbit of the Kremlin were progressively placed under the control of oligarchs or financial interests close to the government. Many leading digital entrepreneurs were expropriated, as was the case for Pavel Dourov, founder of Vkontakte, who was ousted from his own company. Prominent businessmen, such as Alisher Usmanov, took control of entire groups, which thus fell into the hands of interests indirectly controlled by the power networks. In addition, numerous laws were passed to reinforce this control with appropriate technical measures. This was the case with the so-called "Yarovaya" laws (2016), which, by obliging Russian platforms to set up backdoors and share their encryption keys with the authorities, inaugurated a real system of mass surveillance (Sivetc 2021).

In addition to these major political and legislative maneuvers, the digital space will gradually become part of the discursive universe of power and occupy a prominent place in the large family of language elements on which Kremlinocentrism is based. The term Runet, which in the 1990s and 2000s was synonymous with a place of intense creativity, ingenuity, as well as illegality, will gradually change its meaning as it is taken up by official communication. This recuperation goes through many strategies, the most remarkable of which is the one concerning the political role of the Russian language. The Russian language, which is the primary objective characteristic of the Runet, has become a major object of geopolitical tensions in the post-Soviet space. The centrifugal dynamics that oppose Kremlinocentrism in several former republics are based on the (re)construction of national narratives and, above all, the enhancement of national languages. In other words, this is a dynamic of emancipation through nation branding on the part of states that consider themselves freed from the identity, memory and linguistic tutelage exercised by Moscow for many decades. These centrifugal dynamics are generally perceived as threats by Russia because they are assimilated to a desire to free itself from the post-Soviet "community of destiny", its history and its myths, to embrace other models. From then on, the practice of the Russian language takes on an extremely strong identity and geopolitical meaning, as evidenced by the role that the status of the Russian language plays in the Ukrainian conflict (Kulyk 2016). As a Russian-speaking object par excellence, the Runet becomes the receptacle and amplifier of these conflicts – all the more so as the revolutions that overthrow or threaten to overthrow pro-Russian leaders, in Ukraine or Belarus, are organized on the Internet. In the same logic, several governments now consider the Runet platforms as tools of influence under the control of the Kremlin. This is the case, for example, in Ukraine, which has banned the use of Russian platforms on its territory for reasons of national security, or in Estonia, which closely monitors the algorithms of the news aggregator Yandex Novosti: widely used in the Russian-speaking community of the country, this service is suspected of favoring certain information over others in order to influence the representations of the Russian-speaking Estonian community, against a backdrop of latent identity and language conflict.

7.3. Control and dependencies on the lower layers of cyberspace

At the end of the 2010s, the term Runet no longer had the same meaning as 10 years previously. While it used to refer to a kind of digital Wild West, it now refers to an imaginary world and to platforms controlled by the Russian government. This appropriation of digital space has rapidly continued on the lower layers of cyberspace, i.e. not only at the level of the digital dialogue spaces constituted by the platforms, but also at the level of the hardware and technical protocols through which the data circulate that make the existence of such spaces possible. One of the first actions taken by Russia in this direction dates back to 2014, when the Duma adopted Law 242-FZ, which obliges platforms to store data belonging to Russian individuals or legal entities on Russian territory. At the time, this law was much discussed at the global level, as it violated the logic of free flow of data on a network, the Internet, which is supposed to escape borders, territorialities and geography. This law was presented by the Russian government as a measure to protect itself against the massive American eavesdropping that Snowden had revealed a year earlier. It had important consequences for the organization of the post-Soviet network, insofar as the obligation to store data in Russia led to a surge in the demand for data centers, where the data we consult on the Internet are physically stored. In some regions, such as southern Siberia[4], there were many storage facilities where Russian platforms came to store their users' data, whether they were Russian or other post-Soviet nationalities (Estecahandy and Limonier 2020). From then on, we witnessed the emergence of real strategic territories in Russian-speaking cyberspace, where the data of Runet users was stored. In addition to southern Siberia, a large part of the Runet's digital memory is concentrated in the Moscow region, the Leningrad oblast and Karelia.

A few years later, this dynamic of data relocalization was followed by ambitious attempts to control the flows and routings through which data circulates on the

4. Southern Siberia has several advantages: on the one hand, the cold weather that prevails there for part of the year makes it possible to regulate the temperature of servers for free. On the other hand, the price of electricity is very low there, due to the number of hydroelectric dams around Baikal. Finally, South Siberia is well connected to the large fiber optic backbones along the Trans-Siberian corridor.

network. This is the case for the so-called "Sovereign Runet" law, which aims to give the state the technical and legal means to control data packets entering and leaving Russian territory, and to "isolate" the Russian network from the rest of the world in the event of an external threat[5]. In the panoply of measures adopted over the past decade to control the Internet, this law is by far the most ambitious, as it requires an in-depth reorganization of the network and unprecedented coordination between the countless actors on the Russian Internet. Previous laws focused on access to data held by platforms; the 2019 law requires operators and Internet Service Providers to install, on the nodal points of their network, boxes operated by Roskomnadzor (the Russian telecommunications gendarme, equivalent to the French Cnil) called "anti-threat apparatus" (TSPU)[6]. These quasi-digital border posts give the Russian state the right to monitor data packets that cross Russian borders, and, in the event of an external threat, they allow it to disconnect the Russian part of the Internet[7].

Since 2019, it has been understood that the state's strategy is to fully control the data circulating in Russia. In the early days of the Internet, it was believed that cyberspace escaped the geography of power (Virilio 1997), because of its highly distributed architecture, the ultra-high speed of transmissions, the global nature of transmission protocols and the omnipotence of intermediation platforms, which made any state control difficult. With digital border crossings, Russia wants to prove the opposite.

It has certain assets to achieve its ends and to set up these digital borders behind which the Federation as well as former Soviet republics could find themselves entrenched – voluntarily or not. The post-Soviet zone remains dependent on an organization of the Internet network centered around Russia and Moscow. In other words, in many post-Soviet countries, any data packet entering or leaving will necessarily pass through a Russian router or through cables crossing Russian territory. Considering the eavesdropping technologies that are currently being developed within the framework of the Sovereign Runet, Russia would theoretically eventually have the possibility of monitoring Internet traffic passing through its territory, or even filtering or interrupting it (Limonier 2021). At the same time, the possibility of disconnecting the Runet from the rest of the world in the event of an external threat would place certain countries in a delicate position, given their dependence on Russian connectivity.

5. Federal Law No. 90-FZ of May 1, 2019 amends the content of about 10 previous laws on telecommunications regulation. For a complete list, see: https://rg.ru/2019/05/07/fz90-dok.html.

6. Article 46 of Federal Law No. 126-FZ of July 7, 2003, as amended by Law No. 90-FZ.

7. Order No. 127 of February 12, 2020 of the Government of the Russian Federation.

This connectivity is organized around a St. Petersburg–Moscow–Ekaterinburg–Vladivostok axis, which follows the route of the Trans-Siberian Railway. On this axis, powerful cables connect Europe to Asia along three outbound routes: the first connects St. Petersburg to Scandinavia, Germany and Great Britain via a cable system running along the Baltic Sea floor. The second connects Moscow to Frankfurt (Europe's main data exchange hub) via the Moscow–Smolensk–Minsk–Berlin railway axis. The third major corridor is located in the Far East: it consists of several cables linking Vladivostok to Japan, from which many transpacific cables connect Asia to the West Coast of the United States.

Many former Soviet republics depend on these three corridors to connect to the rest of the world. This is primarily due to the fact that the cable network is largely inherited from the rail, road and telephone routes built during the imperial or Soviet era. This results in a situation of dependence on Russia, particularly in Central Asia. The countries of this region can only ensure their connectivity via a north–south axis that follows the major road and rail corridors leading to Russia through the Kazakh steppe. In Central Asia, there are almost no connections to the south and the Afghan or Iranian mountains, while the new Silk Roads planned by China have not yet succeeded in opening up the connectivity of this region on an east–west axis. Some countries such as Kyrgyzstan can even be described as a digital dead end, as they are so dependent on their Kazakh neighbor (which is itself dependent on Russia) for connectivity. The main cables linking this country to the outside world follow the Bishkek–Almaty route, bypassing the Tian Shan. The rest of the country, which is mostly made up of high mountains that are difficult to walk on, would require costly investments.

In other regions, such as the Caucasus, dependence on Russia is also very high, especially in Armenia and Azerbaijan, both of which depend on Russian infrastructure. Georgia, on the other hand, has implemented an interesting emancipation policy in the wake of Russia's invasion of South Ossetia in August 2008. Before that date, Georgia depended on Russian cables running to Sochi (coastal cable) and Vladikavkaz along the Military Route. Today, the country is connected via the Caucasus Cable System, which crosses the Black Sea and connects the Georgian port of Sochi directly to Bulgaria and further to Europe via the Balkans. Interestingly, this Caucasus Cable System was the subject of an attempted takeover in 2016 by a Russian company. The deal was aborted after it was revealed in the press, causing a mini national scandal. This episode, which was highly publicized in Georgia, illustrates the sensitivity of the issue of infrastructural dependencies that frame the Runet in its lower layers.

In the western part of the former USSR, the situation is more mixed. For example, Ukraine, which used to be a digital bridge between Russia and Europe, saw the quality of its connections to Moscow deteriorate after the Mayan revolution and the start of the civil war in the Donbass. In many ways, Ukraine has become a digital frontline between the sphere of influence of the Runet and the global Internet (Pétiniaud and Salamatian 2020). As the Kiev government seeks to emancipate itself from any control exercised by Moscow over its digital networks, Crimea and the Donbass have been progressively attached to Russian networks (Douzet et al. 2020). In Crimea, this has resulted in the destruction of the cables connecting the peninsula to mainland Ukraine via the Perekop Isthmus, and the construction of new corridors linking the region to Russia via the Kerch Strait. In the Donbass, a few Russian-based operators have taken control of the Internet in the separatist territories, while the cables crossing the front line have been destroyed.

Further north, Belarus seems to be well connected to Russia in terms of infrastructure. It is through this country that a large part of the data circulating between Russia and Europe passes, while the highly centralized structure of the network allowed the Internet to be cut off in August 2020, at the time of the demonstrations against the continuation of Alexander Lukashenko's rule.

Finally, the Baltic States are almost completely free of any dependence on Russia for data transit: only two cables link the three Baltic States to Russia, while a dozen undersea backbones firmly link this area to Scandinavia and Germany.

7.4. Conclusion

The Runet constitutes a geopolitical phenomenon in the fullest sense of the term because it is indeed a space (admittedly multidimensional) that has become an issue (political, security, identity, memory), to use the expression coined by Stéphane Rozières. Although the Runet is now the subject of numerous sociological studies (Ermoshina and Musiani 2017), political science or history (Kolozaridi and Muravyov 2020), few works approach it from the angle of geography, and even less from geopolitics. Certainly, many works mobilize a geographical terminology to approach the question of the sovereign Runet, by speaking of "territorialities" or "digital borders" to evoke the dynamics of control and digital censorship that are currently taking place in Russia (Kukkola and Ristolainen 2018). However, few have attempted to undertake a true spatial analysis of the phenomenon, especially in its regional dimension, probably because there is a real difficulty in collecting and analyzing the telemetric data that are indispensable for spatializing the Internet. Producing geopolitics for a segment of cyberspace such as the Runet is therefore a fundamentally multidisciplinary affair, which combines research in computer science and geography (Limonier et al. 2021). It is also a sensitive matter, since the data essential to the spatialization of our object is strategic knowledge, in the sense that Yves Lacoste understands it. Thus, it is not surprising that the research that most closely corresponds to a geopolitical approach to the Runet is conducted by the Institute of Informational Problems at Lomonossov University… which precisely promotes the Russian government's project of control (Pilyugin 2017).

7.5. References

Douzet, F. (2014). La géopolitique pour comprendre le cyberespace. *Hérodote*, 152–153, 3–21.

Douzet, F. (2020). Éditorial. Du cyberespace à la datasphère. Enjeux stratégiques de la révolution numérique. *Hérodote*, 177–178, 3–15.

Douzet, F., Pétiniaud, L., Salamatian, L., Limonier, K., Salamatian, K., Alchus, T. (2020). Measuring the fragmentation of the internet: The case of the border gateway protocol (BGP) during the Ukrainian crisis. *12th International Conference on Cyber Conflict (CyCon)*, 157–182.

Ermoshina, K. and Musiani, F. (2017). Migrating servers, elusive users: Reconfigurations of the Russian internet in the post-Snowden era. *Media and Communication*, 5(1), 42–53 [Online]. Available at: https://www.cogitatiopress.com/mediaandcommunication/article/view/816.

Estecahandy, H. and Limonier, K. (2020). Cryptomonnaies et puissance de calcul : la Sibérie orientale, nouveau territoire stratégique pour la Russie ? *Hérodote*, 177–178, 253–266.

Grumbach, S. and Frénot, S. (2014). Des données à l'intermédiation, une révolution économique et politique. In *Big Data. Nouvelles partitions de l'information*, Calderan, L., Laurent, P., Lowinger, H., Millet, J. (eds). De Boeck Supérieur, Louvain-la-Neuve.

Kolozaridi, P. and Muravyov, D. (2020). The narratives we inherit: The local and global in Tomsk's internet history. *Internet Histories*, 4, 49–65.

Kukkola, J. and Ristolainen, M. (2018). Projected territoriality: A case study of the infrastructure of Russian digital borders. *Journal of Information Warfare*, 17, 83–100.

Kulyk, V. (2016). Language and identity in Ukraine after Euromaidan. *Thesis Eleven*, 136(1), 90–106.

Lacoste, Y. (2014). *La géographie, ça sert, d'abord, à faire la guerre*. La Découverte, Paris.

Limonier, K. (2021). Vers un "Runet souverain" ? Perspectives et limites de la stratégie russe de contrôle de l'Internet. *Echogeo*, 56 [Online]. Available at: https://journals.openedition.org/echogeo/21804#:~:text=28Le%20premier%20et%20principal,Russie%20selon%20le%20RIPE10.

Limonier, K., Douzet, F., Pétiniaud, L., Salamatian, L., Salamatian, K. (2021). Mapping the routes of the Internet for geopolitics: The case of Eastern Ukraine. *First Monday*, 26(5) [Online]. Available at: https://firstmonday.org/ojs/index.php/fm/article/view/11700.

Mareï, N. and Richard, Y. (eds) (2018). *Dictionnaire de la régionalisation du monde*. Atlande, Neuilly.

Pétiniaud, L. and Salamatian, L. (2020). Le rôle de la topologie d'Internet dans les territoires en conflit en Ukraine, une approche géopolitique du routage des données. *L'Espace Politique*, 40 [Online]. Available at: https://journals.openedition.org/espacepolitique/8031.

Pilyugin, P. (2017). Problemy opredeleniia granits v informatsionnom prostranstve. *T-Comm: telekommunikatsiia i transport*, 11(8), 37–44.

Raffestin, C. (1980). *Pour une géographie du pouvoir*. Librairies techniques, Paris.

Raviot, J.-R. (2017). Le prétorianisme russe : l'exercice du pouvoir selon Vladimir Poutine. *Hérodote*, 166–167, 9–22.

Sivetc, L. (2021). Controlling free expression "by infrastructure" in the Russian Internet: The consequences of RuNet sovereignization. *First Monday*, 26(5) [Online]. Available at: https://firstmonday.org/ojs/index.php/fm/article/view/11698.

Ventre, D. (2011). *Cyberespace et acteurs du cyberconflit*. Hermès Science Publications, Paris.

Virilio, P. (1997). Fin de l'histoire, ou fin de la géographie ? Un monde surexposé. *Le Monde diplomatique* [Online]. Available at: https://www.monde-diplomatique.fr/1997/08/VIRILIO/4878.

8

Security Regions: A Heterogeneous World Between Conflicts and Cooperation

Emmanuel CHAUVIN
UMR LISST, Université Toulouse Jean Jaurès, France

8.1. Introduction

The region has become a new territory of security, following two recompositions of the field of security. On the one hand, security, as a set of protection, control and regulation devices for individuals, no longer has the state as its exclusive subject (the one who secures), but various political levels. On the other hand, the object of security (what needs to be secured) is no longer only the defense of the integrity of the state: it is a question of preventing all threats to the lives of populations, in a field that goes from the classical domains of security (defense, diplomacy, justice, police) to those of human security (food, energy, environment, etc.) (Gros 2012).

Security regions in the world are groups of contiguous spatial units that are dependent on each other through strong security interactions. They can take the form of security communities, characterized by their peaceful coexistence (Deutsch 1957; Adler and Barnett 1998), or the form of regional conflict complexes (Wallensteen and Sollenberg 1998; Buzan and Wæver 2003). The construction of a security community requires the strengthening of regional cooperation to reduce threats (regionalization of security), while the regional conflict complex is constructed through the increase in cross-border violence and conflicts (regionalization of insecurity).

Regionalization of the World,
coordinated by Pierre BECKOUCHE and Yann RICHARD.

However, can we divide the world into regions where the regionalization of security and security communities dominate, and regions marked by the regionalization of insecurity and regional conflict complexes? On the one hand, the anglophone political science literature has theorized the subject of regional security complexes. Within this framework, authors have proposed a classification of the world's regions in a spectrum ranging from security communities to conflict formations (Buzan and Wæver 2003). On the other hand, French geography has never made the link between security and region a theoretical problem, approaching security at the regional scale through a succession of case studies and through privileged spatial objects, such as borders (Sintès 2015; Blondel 2016; Roussel 2016; Szary and Cattaruzza 2017; Bennafla and Meier 2018), transnational water zones (Raison and Magrin 2009; Bouquet 2011; Blanchon 2019) and forced migration (Lassailly-Jacob and Quesnel 1999; Cambrézy 2001; Clochard 2007; Chauvin 2015; Vignal 2021). We will weave links between these two fields of research, with a tropism towards Africa linked to our own research.

The first step will be to study the construction of regional security architectures *through* security cooperation. These architectures have diversified both in their locations and in their modes of operation. Secondly, we will analyze the role of insecurity in the formation of regions. Conflicts and violence are structured on a regional scale and recompose institutional and functional regions. However, insecurities presented as regional only have empirical realities at the local or national level, and this overvaluation of the extent of threats serves security policies.

8.2. Regional security cooperation: globalization and diversification

Security has been the basis of many regional organizations: guaranteeing peace between neighboring states and influencing rivalries between world powers have been powerful drivers of regional construction. The need for security cooperation has given rise to the creation of security architectures, following a double movement of globalization, with security regions developing all over the world, and diversification of their objects and modes of operation.

8.2.1. *Security, the basis of regional construction*

8.2.1.1. *Peace as an objective of regional construction*

Europe is the archetype of a region built with the objective of peace, right from its beginnings. The idea of Europe, as a political construction between states sharing common institutions, appeared in the 17th century when the new art of governing no longer had as its main object the imperial dream or Christian universalism, but peace between neighboring states or the balance of their forces, which presupposed the constitution of security mechanisms (Foucault 2004). After three major wars in less than a century, the creation of the European Coal and Steel Community (1951) was to make "any war between France and Germany [...] not only unthinkable, but materially impossible", by pooling production related to the arms industry (Schuman 1950).

After decolonization, the consolidation of the internal and external security of states was also one of the objectives of the creation of regional organizations in the Third World. Thus, the Organization of African Unity (OAU) promulgated the intangibility of borders in 1964, with the objective of avoiding inter-state border conflicts and internal rebellion movements that could lead to secession. Another example is the creation of the Association of Southeast Asian Nations (ASEAN), which was motivated by the need to bring together countries with deep political disagreements that considered their security threatened (Malaysia, Indonesia, Singapore) (Braillard and Djalili 1985).

8.2.1.2. *The region as a geostrategic tool*

In the context of the Cold War, the creation of regions was used as a geostrategic tool in the rivalries between world powers. The confrontation between the United States and the Union of Soviet Socialist Republics (USSR) gave rise to the multiplication of security alliances to build one's own bloc or to weaken and undermine that of the adversary. Examples include the North Atlantic Treaty Organization (NATO), the Warsaw Pact, the South East Asia Treaty Organization (SEATO), the Baghdad Pact, the Australia, New Zealand, United States Security Treaty (ANZUS) and the Inter-American Treaty on Mutual Assistance.

The end of the Cold War paved the way for an empowerment of regional security dynamics. From the 1990s onwards, new political constructs emerged, and regional organizations were able to develop their own security policies (Buzan and Wæver 2003). Nevertheless, the use of regions as a geostrategic tool by the great powers has persisted. Thus, NATO, a vector of influence for the United States, has undergone a double geographical expansion by intervening outside the treaty area (peacekeeping in the Balkans, war against the Taliban in Afghanistan) and by the integration of new members in Central and Eastern Europe (Cattaruzza 2014). In parallel, Russia created the Collective Security Treaty Organization (CSTO) and China the Shanghai Group (1996), now the Shanghai Cooperation Organization (SCO).

China's assertiveness on the Asian and global scenes is now changing some of this. Security regions are once again clearly becoming major geostrategic tools, like the Indo-Pacific security alliance led by the United States to counter Chinese influence (Quadrilateral Security Dialogue (QUAD), Aukus). Through its "New Silk Roads", Beijing is adopting, in addition to its influence in regional organizations, a reticular approach mixing economic influence and military capabilities (Lasserre et al. 2019).

8.2.2. *Security for the regions: from defense to human security*

8.2.2.1. *The growing role of the regions in defense*

One of the markers of the growing involvement of regional organizations in conventional security is the deployment of regional military forces in theaters of war. NATO, the European Union (EU) and African regional organizations are the most active in this area, either alone or in the framework of the United Nations (UN). Since the 1990s, the UN has indeed multiplied peacekeeping operations in partnership with regional organizations (Liégeois 2017).

In Africa, for example, regional military coalitions have been numerous since the 1990s. From 1990 to 2017, there have been 42 UN peace operations in Africa, most with African forces involved, including 16 from the African Union (AU), seven from the Economic Community of West African States (ECOWAS), two from the Intergovernmental Authority on Development (IGAD) and one from the Community of Sahel-Saharan States (CEN-SAD), the Economic and Monetary Community of Central Africa (CEMAC), the Economic Community of Central African States (ECCAS) and the Southern African Development Community (SADC) (Chauvin 2018a) (see Figure 8.1). African states find military experience, funding, and regional projection capabilities. Western states, such as France and the United States, support and fund these regional peace operations because they see them as a tool for influence without direct military engagement, the latter often being costly and unpopular.

Because of the operational inefficiency and, at times, the bias of the troops involved, the model of intervention by regional troops under a UN mandate is often challenged, leading to the use of alternatives. Smaller African regional military powers (South Africa, Uganda, Chad, Rwanda, etc.) intervene directly in conflicts, with or without Western support. Russia also proposes to intervene – notably through Wagner's mercenaries, a largely asymmetric military instrument (guerrilla warfare, sabotage, disinformation, etc.) – to extend its political and economic influence (mines, hydrocarbons). In countries experiencing conflicts or security crises, especially when regional forces have become bogged down by their inaction (operational weakness, limited mandate, etc.), Wagner's brutal methods against insurgents can seduce governments (Central African Republic, Libya, Mali, Mozambique, Democratic Republic of Congo – DRC – Sudan) (Bertrand 2020).

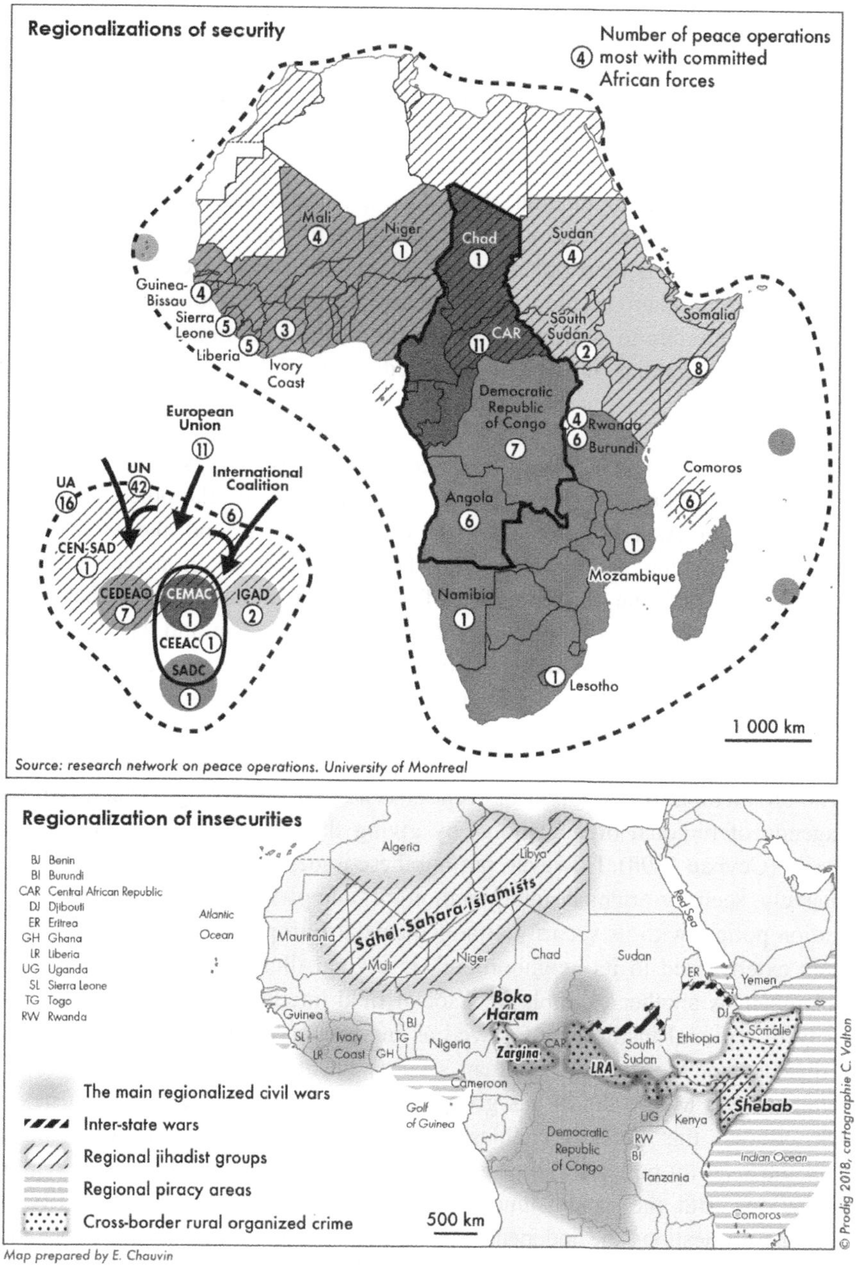

Figure 8.1. *Regionalizations of security and insecurity in Sub-Saharan Africa (1990–2017) (source: Chauvin 2018a)*

8.2.2.2. *Human security, the new driving force of regional integration?*

Human security is an important political object in regional security communities. Some authors consider that in these communities, the actors stop seeing themselves as threats and therefore carry out a "de-securitization", i.e. an exit from security policies, in order to turn to other subjects of cooperation (Buzan and Wæver 2003). More realistically, we might think that, since the traditional fields of security are already the subject of cooperation, states are turning to other fields of security. Thus, the European Union is developing security policies in many areas, both internally and in its external relations (food, energy, maritime, etc.). Promoting human security also makes it possible to appropriate prerogatives that were previously reserved for states without becoming too overtly involved in areas that are still considered as sovereign.

Human security, which appears to be consensual, also serves to divert attention from military issues and rivalries between states in regional relations. China is thus implementing in Asia a "diplomacy of cooperation" around the environment, hydrocarbons, fishing and science. This favors a decrease in perceived threats between states and the development of multi-domain cooperation that allows military-political issues to be circumvented or appeased (Fau and Tréglodé 2018). But the soft power and multilateralism proposed by China as part of this "rim diplomacy" do not preclude an aggressive defense policy against Taiwan and American influence in Maritime Asia (Cabestan 2007).

The broadening of the notion of security is a lever for putting societal issues on the agenda of regional organizations, by giving them a character of urgency and necessity (Ceyhan 1998). However, putting these issues on the agenda can lead to an exclusively security-oriented interpretation of complex social issues. The EU's migration policy, with its security arrangements at Europe's external borders and at borders externalized to its neighborhood, is a good illustration (Migreurop 2012). Development is another example of a social field interpreted through the prism of security in regional institutions (as well as in states and international institutions). The diffusion of the notion of human security and the theory of the fragile state (Châtaigner and Magro 2007) has given rise to the "security-development nexus", which assumes that security enables development and vice versa. This "nexus" infuses the security policies and programs of the development agencies of regional organizations. But the establishment of a systematic link between security and development questions the independence of development aid, which may be more oriented towards the security interests of Northern regions and donor states than towards the development problems of Southern recipient countries (Amer et al. 2012).

8.2.3. *Regional security architectures: a variety of styles*

8.2.3.1. *Overlapping multi- and minilateral security architectures*

Multilateral security architectures vary in size. Super-security regions, such as NATO or the OSCE (Organization for Security and Cooperation in Europe), are organizations that span continents and bring together political and functional regions. Regional organizations of continental or intra-continental size also have security policies, whether they are multi-domain organizations (European Union, African Union, Union of South American Nations – UNASUR – etc.) or organizations more specialized in security (Gulf Cooperation Council, for example).

Minilateralism in security, cooperation among a small number of states, is often an alternative to the inefficiency of multilateral security architectures. It offers more flexibility and opportunities to reach agreements because the number of parties is smaller. It is true that security regions have varying levels of operationality. The African Union is a paper tiger, with security arrangements that reflect its own image, i.e. not very operational (African Peace and Security Architecture, African Standby Force). At the other end of the spectrum, the European Union's policies are more robust, both in human and conventional security, even though the results are sometimes more limited than expected, as in the field of defense. But security minilateralism can also be seen as "club diplomacy", a cooperation between the most powerful states, which excludes other states from decision-making (Group of Seven – G7 – for example) (Badie 2011).

8.2.3.2. *Diversity of operation of regional security architectures*

Regional security architectures are influenced by their external relations with other regions, states, transnational actors and intergovernmental actors. One example is the influence of security regions from the North on the South in terms of the circulation of regulations and norms (Bagayoko-Penone and Hours 2005), funding and training, as can be observed in the case of the European Union in Africa in the military field. In addition to numerous military and civilian missions conducted in Africa under the Common Security and Defense Policy (CSDP) (23 in Africa out of 42 deployed worldwide since 2000), the EU finances the logistics of African Union peacekeeping operations through development funds ("African Peace Facility"). Since 2021, it can directly finance sub-regional or state military initiatives without going through the AU ("European Peace Facility"), as well as broaden the targets of funding beyond logistics (financing of military equipment, including lethal, military training, etc.). As another example, the European Union is funding and contributing to the organization of the G5 Sahel Defense College, a war school for senior officers from Sahelian countries, through the "G5 Sahel Support Project for Sahel Security", which is itself funded by the "European Union Emergency Trust Fund for Africa". Thus, part of the regional military organization in Africa depends on the EU and its member states.

The functioning of regional security architectures also varies according to the relations between member states. The intensity of security cooperation depends largely on the degree of cooperation or rivalry between them. Similarly, the orientations and models of regionalization vary the regionalization of security. As an example, the divergent strategies of the many regional organizations in South America or the opposing models of regionalization in Central Asia (integration, bilateralism, neutrality, etc.) (Mareï and Richard 2018) hinder the possibility of unified regional security cooperation in these parts of the world. Authors also distinguish security regions according to the greater or lesser symmetry of inter-state relations within them: polarized regions, dominated by some state(s) or group(s) of states (North America, CIS, EU-Europe, East Asia) with hegemonic tendencies can in this respect be distinguished from standard regions where asymmetries between states are not very marked (South America, Middle East, Africa, South Asia) (Buzan and Wæver 2003).

The functioning of states also has an impact on the regionalization of security. Ideologies can promote security, or hinder it, for example, in states where nationalism or sovereignty prevails. Moreover, state dysfunction percolates into regional security practices. For example, indiscipline, clientelist networks and the privatization of the state in national security forces in Africa permeate the functioning of African contingents of regional peacekeeping troops deployed on the continent.

Thus, regional security cooperation, which is part of regional construction, has become globalized as the world has become more regionalized. At the same time, from defense to human security, the fields of regional security cooperation have diversified, giving rise to very different regional security architectures, both in terms of size and in terms of their operating methods. But the multiplication of security architectures does not prevent the development of regional insecurities.

8.3. Realities and myths of the regionalization of insecurity

Conflicts and violence, through their cross-border extension, can lead to the construction of regional formations that link contiguous spaces through shared threats. Whether or not they extend regionally, violence and conflict recompose functional and institutional regions *through* countervailing movements of regionalization/de-regionalization. However, the very idea of a regional architecture of insecurity is sometimes a misinterpretation of empirical reality. It can be a simple discursive construction, an imaginary one possibly supported by a cartographic medium, not without political implications.

8.3.1. *Regional violence training*

8.3.1.1. *Regional conflict complexes*

Conflicts between contiguous states are the source of regional conflict complexes. Observing the repetition of certain conflicts, some authors consider that large regions are perennial conflict complexes, such as the Middle East, marked by clashes over borders, hydrocarbons, water, ideologies and cultural affiliations, or South Asia, also over borders and water, trade, migration and nuclear issues (Buzan and Wæver 2003). Nevertheless, this statement is debatable: it proceeds by aggregating a multitude of shifting inter-state conflicts. In these regions, it is on a finer scale that certain localized conflicts are more likely to be part of permanent regional formations, such as the Israeli–Palestinian conflict in the Middle East or the conflicts between India and Pakistan in South Asia.

Other regional conflict formations arise from the regionalization of civil war(s) in three main configurations:

– the conflict system, i.e. the regional intertwining of two neighboring and originally distinct civil wars: this was the case between Liberia and Sierra Leone in the 1990s (Marchal et al. 2002), or between Chad and Sudan (Darfur) in the 2000s (Marchal 2006), for example;

– regional conflict escalation, intervention of regional actors in an intra-state conflict without the clashes spilling over beyond the borders of the state at war, as in the conflict in the Central African Republic, where a number of fighters from Central Africa (Chad, Sudan, Congo, etc.) intervene without spreading the fighting to neighboring countries (Chauvin 2018b);

– regional conflict contagion, i.e. the spread of a civil war to an adjacent area, as in the mid-1990s from Rwanda to the rest of the Great Lakes and the DRC (Pourtier 2000).

8.3.1.2. *Transnational violent networks*

Other transnational threats do not necessarily lead to conflict, but they do lead to violence. Some criminal activities can be deployed on a regional scale. In this case, the region can be an element and an echelon in a global criminal archipelago (drug trafficking, cybercrime, wildlife poaching, diamond trafficking, etc.). Intense links then develop between the main production basins and transit locations of these globalized criminalities: for example, for heroin and cocaine narcotrafficking around the Golden Crescent and Colombia as production basins and West Africa as one of the transit hubs to Europe (Magrin et al. 2018). Other criminal activities are even more regionalized: piracy in various areas of the globe (Strait of Malacca, Red Sea/Indian Ocean, Gulf of Guinea, etc.) and kidnapping for ransom by highway robbers in Central Africa (Central African Republic, Cameroon, Chad, Nigeria) (Issa 2010), for example. In the latter case, gangs organize themselves into cross-border networks linking residential bases, located on the borders of states, around which mass kidnappings are carried out (Chauvin and Seignobos 2013).

Armed jihadist groups also play with borders and often organize themselves at a regional level, despite international affiliations that are sometimes simply a sign (to Daech or al-Qaeda). Here again, Africa is full of examples: in the western part of the Sahel-Sahara (Mali, Niger, Burkina Faso), around Lake Chad (Niger, Nigeria, Cameroon, Chad) and in part of the Horn of Africa (Somalia, Kenya, Uganda). Boko Haram in the Lake Chad region occupies space in the same way as the coupeurs de route: the Islamist movement does not seek to reproduce a continuous and hierarchical state grid, but favors a reticular logic, circulating from mobile camps and border sanctuaries (Nigeria, Cameroon, Niger, Chad (Seignobos 2019)).

8.3.2. *Insecurities that recompose regions*

8.3.2.1. *Insecurity as a factor of de-regionalization*

Insecurity is a powerful factor in functional de-regionalization, reducing cross-border flows (economic, migratory, etc.). Conflicts often lead to regional economic isolation, reducing flows between the conflict theater and neighboring areas. In Central Africa, for example, the Central African Republic is a model of a neglected periphery that has become a blind spot for regionalization because of armed violence. Similarly, the Lake Chad region, which used to be a functional regional trading system, has been neglected as a result of a crisis, the establishment of Boko Haram, and the counterinsurgency policies of the states bordering the lake (Chauvin and Magrin 2020). Neighboring spaces often benefit from the detour of flows from insecure areas, redrawing the geography of functional regionalization.

Regional conflicts can also block political regionalization between two countries (Chad vs. Sudan, DRC vs. Rwanda and Uganda (Liégois 2017)), or more broadly, across the region. To a large extent, tensions between North African states block regionalization in this part of the world. Another example is the current political crisis in Venezuela, with armed violence against civilians, which maintains and reinforces the crisis of regional integration in Latin America, undermining an integration already weakened by its profusion and politicization. The number of regional organizations is high in this continent, and these organizations are segmented between a progressive and a neoliberal tendency, with both tendencies clashing over the recognition or rejection of Nicolas Maduro's Venezuelan power (Gaudichaud and Posado 2021).

8.3.2.2. *Regionalization through violence*

According to another point of view, violence can promote functional regionalization, by generating flows linked to military activities (combatants, weapons, etc.), the war economy (trafficking of all kinds, humanitarian aid, etc.) or violence against civilians (refugees). Thus, for example, cross-border migratory spaces are appearing around states at war (Syria, Venezuela, Afghanistan, various African states, etc.), linked to forced migrations and post-migratory circulations around border refugee camps.

Insecurities can also give rise to new inter-state security cooperation to deal with transnational threats, such as within ASEAN in the face of maritime piracy in the Strait of Malacca (Frécon 2020) or in Africa in the face of the jihadist threat. In the latter case, two regional security architectures bringing together countries belonging to different regional institutions have been created: the Multinational Joint Task Force (MMF) against Boko Haram (Benin, Cameroon, Nigeria, Niger, Chad) and the G5 Sahel against jihadist movements in the Sahel (Burkina Faso, Mauritania, Mali, Niger, Chad). Insecurities have forced states to go beyond their usual regional institutional frameworks (West Africa, on the one hand, Central Africa on the other), although there is no certainty about the sustainability of these two security architectures and their impact on regional integration beyond the security field.

8.3.3. *Imaginary figures of regional insecurities*

Regional architectures of insecurity are sometimes purely intellectual constructs, imaginations created by the misrepresentation of threats on a regional scale when they only have empirical realities on a local or national scale. The notions of "crisis crescent" or "crisis arc" or "instability", "region" or "unstable zone", which are used differently depending on the country (Foucher 2012), often overestimate the phenomena of insecurity, by aggregating distinct elements without there necessarily being interactions between them and a real regional dynamic. From this perspective, the Sahel is often wrongly described as a region whose homogeneity is due to the jihadist threat, even to the point of using the term Sahelistan, a neologism that has been strongly criticized for its "decontextualizing" analogy (Afghanistan vs. the

Sahel), its lack of precision in the delimitations (the entire Sahel is far from being the theater of the Islamists), and its excessive generalization (Bonnecase and Brachet 2013).

The construction of imaginary figures of regional insecurity can be achieved through cartography. Some institutional or media cartographies tend to overextend the geographical extension of security crises, simplifying them for political reasons or in order to reduce the risks incurred by people who wish to travel there. This can be seen in the maps of the Ministry of Foreign Affairs in France, which reflect a concern for the protection of its French nationals abroad (Bouquet 2012).

Political discourses turn the preservation of identity and the defense of culture into a regional security issue by artificially pitting regional blocs against each other. One of the best known examples is that of Samuel Huntington, who identified regional groupings on a primarily cultural basis, with the idea of an inevitable confrontation between the West and Islam. Various French geographers have endeavored to deconstruct these discourses, either implicitly, by showing the ethno-centric and historical character of the divisions of the world (Grataloup and Capdepuy 2013), or explicitly, by showing that the world could be divided up in popular representations according to logics other than the clash between civilization areas (Didelon Loiseau and Richard 2020).

8.4. Conclusion

To some extent, Buzan and Wæver's (2003) regionalization of the world might still apply: Africa, South Asia, the Middle East apprehended as conflict formations, North America and Europe as security communities, ASEAN and the Southern Cone of Latin America as security regimes (conflicts but cooperation to deal with them), the former Soviet republics and East Asia as mixtures between security regimes and conflict formations. However, this division of the world into security and insecurity blocs is being challenged by the globalization of regional security architectures: security cooperation has developed everywhere, including in parts of the world where conflicts are numerous, such as Africa, implying a stronger co-presence between conflict formations and security regimes. Moreover, this division, which postulates that in security communities we are gradually witnessing a reduction in security policies in favor of other subjects of cooperation, underestimates the considerable enlargement of the notion of security, the corollary of which is the multiplication of mechanisms for the protection and control of people in various fields (health, environment, energy, etc.) rather than a reduction in the scope of security. Finally, the impression that emerges from various case studies is that regional security architectures are often built on continental scales or within large regions, whereas the formations of violence and regional conflict complexes generally concern only a few neighboring countries, on the scale of small regions rather than on the scale of continent-regions.

Isn't the regionalization of security now being called into question? Regional security architectures still appear to be an essential link in the chain of peace between neighboring states, and security cooperation is sometimes a sufficiently consensual tool to promote regional construction. But phenomena such as jihadism and regionalized or globalized crime are building reticular spaces that play on the borders of regional security organizations. And the partly reticular strategy of the Chinese "New Silk Roads" and the Russian asymmetry of the Wagner enterprise implanted by sprawl in developing countries and in states in crisis are fracturing regional security organizations or recomposing regional security.

Regional security is a marginal object of study in francophone geography for several reasons. On the one hand, unlike political science in particular, geography favors an empirical approach based on field surveys. This implies, for the study of regional security dynamics, multiplying the number of study sites on both sides of borders, in regions often made difficult to access by violence. Most studies on security are thus limited to a country or a sub-national region, and thus restricted to one side of the border. On the other hand, there is no branch in French-speaking geography devoted to the study of security, peace, war or violence, and little editorial gathering on the subject among specialist researchers, unlike in English-language geography (see, among others, Flint 2004; Gregory and Pred 2006; Kobayashi 2009). Nevertheless, these themes seem to be increasingly present as objects of scientific study, well beyond the debates on the usefulness of the discipline for making war or making peace.

8.5. References

Adler, E. and Barnett, M. (1998). *Security Communities*. Cambridge University Press, Cambridge.

Amer, R., Swain, A., Öjendal, J. (2012). *The Security-Development Nexus. Peace, Conflict and Development*. Anthem Press, London.

Badie, B. (2011). *La diplomatie de connivence. Les dérives oligarchiques du système international*. La Découverte, Paris.

Bagayoko-Penone, N. and Hours, B. (2005). *États, ONG et production des normes sécuritaires dans les pays du Sud*. L'Harmattan, Paris.

Bennafla, K. and Meier, D. (2018). Enjeux frontaliers et territoriaux au Sahel et au Moyen-Orient. *Égypte/Monde arabe*, 2(18).

Bertrand, M. (2020). Wagner, société militaire privée (SMP) et nouvel outil de la stratégie géopolitique moderne russe. *Sécurité globale*, 24(4), 43–66.

Blanchon, D. (2019). *Géopolitique de l'eau. Entre conflits et coopérations*. Le Cavalier Bleu, Paris.

Blondel, C. (2016). Aménager les frontières des périphéries européennes : la frontière Serbie/Croatie à l'épreuve des injonctions à la coopération et à la réconciliation. PhD Thesis, Université de Tours, Tours.

Bonnecase, V. and Brachet, J. (2013). Les "crises sahéliennes" entre perceptions locales et gestions internationales. *Politique africaine*, 130(2), 5–22.

Bouquet, C. (2011). Conflits et risques de conflits liés à l'eau en Afrique. *Les Cahiers d'Outre-Mer*, 255, 341–362.

Bouquet, C. (2012). Comment le Nord cartographie les risques au Sud : "Conseils aux voyageurs" et "Risques-Pays" en Afrique. *Bulletin de l'association de géographes français*, 89(1), 94–103.

Braillard, P. and Djalili, M.-R. (1985). Les organisations internationales du Tiers Monde : vers l'élaboration d'un nouveau cadre d'analyse. *Études internationales*, 16(3), 493–504.

Buzan, B. and Wæver, O. (2003). *Regions and Powers. The Structure of International Security*. Cambridge University Press, Cambridge.

Cabestan, J.-P. (2007). La politique étrangère chinoise : une Chine sans ennemis n'est pas forcément une Chine rassurante. *Hérodote*, 125(2), 11–27.

Cambrézy, L. (2001). *Réfugiés et exilés. Crise des sociétés, crise des territoires*. Éditions des archives contemporaines, Paris.

Cattaruzza, A. (2014). *Atlas des guerres et conflits. Un tour du monde géopolitique*. Autrement, Paris.

Ceyhan, A. (1998). Analyser la sécurité : Dillon, Wæver, Williams et les autres. *Cultures et conflits*, 31–32.

Châtaignier, J.-M. and Magro, H. (2007). *États et sociétés fragiles. Entre conflits, reconstruction et développement*. Karthala, Paris.

Chauvin, E. (2015). Violences en Centrafrique : pouvoirs de déplacer, manières de migrer. Centrafricains déplacés et réfugiés (Cameroun, Tchad). PhD Thesis, Université Paris 1 Panthéon-Sorbonne, Paris.

Chauvin, E. (2018a). Afrique subsaharienne. Régionalisation de la sécurité et de l'insécurité. In *Dictionnaire de la régionalisation du monde*, Mareï, N. and Richard, Y. (eds). Atlande, Neuilly-sur-Seine.

Chauvin, E. (2018b). La guerre en Centrafrique à l'ombre du Tchad. Une escalade conflictuelle régionale. Agence française de développement, Paris.

Chauvin, E. and Magrin, G. (2020). Violences et régionalisation en Afrique centrale. *Belgeo*, 4.

Chauvin, E. and Seignobos, C. (2013). L'imbroglio centrafricain. État, rebelles, bandits. *Afrique contemporaine*, 248, 119–148.

Clochard, O. (2007). Le jeu des frontières dans l'accès au statut de réfugié : une géographie des politiques européennes d'asile et de migration. PhD Thesis, Université de Poitiers, Poitiers.

Deutsch, K.W., Burrell, S.A., Kann, R.A., Lee Jr, M., Lichterman, M., Lindgren, R.E., Leowenheim, F.L., Van Wagenen, R.W. (1957). *Political Community and the North Atlantic Area: International Organization in the Light of Historical Experience*. Princeton University Press, Princeton.

Didelon Loiseau, C. and Richard, Y. (2020). La géographie à la recherche des civilisations de Huntington, analyse des représentations du monde. *Bulletin de la société géographique de Liège*, 74.

Fau, N. and de Tréglodé, B. (2018). *Mers d'Asie du Sud-Est. Coopérations, intégration et sécurité*. CNRS, Paris.

Flint, C. (2004). *The Geography of War and Peace: From Death Camps to Diplomats*. Oxford University Press, New York.

Frécon, E. (2020). L'ASEAN face à la redéfinition de sa centralité diplomatique, opérationnelle et géographique. *Hérodote*, 176, 9–24.

Foucault, M. (2004). *Sécurité, territoire, population. Cours au Collège de France (1977–1978)*. Le Seuil, Paris.

Foucher, M. (2012). L'Arc de crise, approche française des conflits. *Bulletin de l'association de géographes français*, 89(1), 6–17.

Gaudichaud, F. and Posado, T. (2021). Le Venezuela, épicentre des tensions géopolitiques latino-américaines. *Diplomatie*, 110, 17–21.

Grataloup, C. and Capdepuy, V. (2013). Continents et océans : le pavage européen du globe. *Monde(s)*, 3(1), 29–51.

Gregory, D. and Pred, A. (2006). *Violent Geographies. Fear, Terror, and Political Violence*. Routledge, New York.

Gros, F. (2012). *Le principe sécurité*. Gallimard, Paris.

Issa, S. (2010). *Les coupeurs de route. Histoire du banditisme rural et transfrontalier dans le bassin du lac Tchad*. Karthala, Paris.

Kobayashi, A. (2009). Geographies of peace and armed conflict: Introduction. *Annals of the American Association of Geographers*, 99(5), 819–826.

Lassailly-Jacob, V. and Quesnel, A. (1999). *Déplacés et réfugiés : la mobilité sous contrainte.* IRD, Paris.

Lasserre, F., Mottet, E., Courmont, B. (2019). *Les nouvelles routes de la soie. Géopolitique d'un grand projet chinois*. Presses de l'Université du Québec, Quebec.

Liégeois, M. (2017). Le rôle des organisations régionales dans le maintien de la paix et de la sécurité internationales. In *La sécurité internationale après Lisbonne. Nouvelles pratiques dans l'Union européenne*, Liégeois, M. and Balzacq, T. (eds). Presses universitaires de Louvain, Louvain-la-Neuve.

Magrin, G., Dubresson, A., Ninot, O. (2018). *Atlas de l'Afrique. Un continent émergent ?* Autrement, Paris.

Marchal, R. (2006). Tchad/Darfour : vers un système de conflits. *Politique africaine*, 102, 135–154.

Marchal, R., Ero, C., Ferme, M. (2002). Liberia, Sierra Léone et Guinée : une guerre sans frontières ? *Politique africaine*, 88, 5–12.

Mareï, N. and Richard, Y. (2018). *Dictionnaire de la régionalisation du monde*. Atlande, Neuilly-sur-Seine.

Migreurop (2017). *Atlas des migrants en Europe. Approches critiques des politiques migratoires*. Armand Colin, Paris.

Pourtier, R. (2000). Guerre et géographie. Du conflit des Grands Lacs à l'embrasement de l'Afrique centrale. In *Politiques et dynamiques territoriales dans les pays du Sud*, Chaléard, J.-L. and Pourtier, R. (eds). Publications de la Sorbonne, Paris.

Raison, J.-P. and Magrin, G. (2009). *Des fleuves entre conflits et compromis. Essais d'hydropolitique africaine*. Karthala, Paris.

Roussel, C. (2016). La frontière syro-jordanienne dans le conflit syrien : enjeux sécuritaires, gestion frontalière. *L'espace politique*, 27.

Schuman, R. (1950). Déclaration du 9 mai 1950. Fondation Robert Schuman [Online]. Available at: https://www.robert-schuman.eu/fr/declaration-du-9-mai-1950.

Seignobos, C. (2019). La région du lac Tchad sous l'hypothèque Boko Haram. *Hérodote*, 172, 63–86.

Sintès, P. (2015). Conflits sans fin à la frontière gréco-albanaise ? *Hérodote*, 158(3), 76–92.

Szary, A.-L. and Cattaruzza, A. (2017). Frontières de guerre, frontières de paix : nouvelles explorations des espaces et temporalités des conflits. *L'espace politique*, 33.

Vignal, L. (2021). Mondialisation, régionalisation, conflits. Comprendre le monde à partir d'un Moyen-Orient en ébullition. Research paper, Université Jean Moulin Lyon III, Lyon.

Wallensteen, P. and Sollenberg, M. (1998). Armed conflict and regional conflict complexes, 1989–1997. *Journal of Peace Research*, 35(5), 621–634.

9

African Integration in All Its Forms

Géraud MAGRIN[1] and Olivier NINOT[2]
[1] *UMR Prodig, Université Paris 1 Panthéon-Sorbonne, France*
[2] *CNRS, UMR Prodig, Aubervilliers, France*

9.1. Introduction

The African continent gives an outward impression of geographic unity – its contours are not debated – which contrasts with the weakness of its regional integration. In 2018, only 15% of its trade was internal and the map of African regional organizations resembled a hodgepodge (Figure 9.1). Yet Africa has not remained aloof to the regional integration dynamics observed in other parts of the world. Since colonization, from the beginning of the 20th century until present, a wide variety of forms of institutional cooperation (political, economic, commercial, monetary, security, water system, environmental management, energy networks) have emerged. However, the regional picture of the continent remains that of an unfinished composition marked by complexity, the overlapping of institutional frameworks (Bach 1998; Magrin et al. 2018), and the weakness of the implementation of agreements, whose effective imprint on societies remains tenuous.

A double tension characterizes African regional integration. On the one hand, the dynamics of modern integration "from above", i.e. through the official institutions backed by the states that emerged from independence, coexist as best they can with relations "from below", cross-border, embedded in the long history, which correspond much more concretely and intimately to the territorial experience of the populations (Igué 1995). On the other hand, the political balkanization that resulted

Regionalization of the World,
coordinated by Pierre BECKOUCHE and Yann RICHARD.

from the decomposition of colonial empires (Besnoit 1979) goes hand in hand with a profound extraversion, also rooted in history (Bayart 2006), whose horizons have become more complex with the latest stages of globalization (Magrin 2013). Here, relations with emerging countries, and China in particular, are superimposed on the web of pre-existing links polarized by Europe and the United States.

The African regional integration scene has been in turmoil since 2015, with the concretization of several regional treaties and the signing of an agreement for a Continental Free Trade Area (CFTA) in 2018. Should we interpret this as the maturation of institutional arrangements finally in touch with endogenous territorial realities, or as yet another vogue responding to new geopolitical conditions marked by the opposition between old western powers and emerging China?

The aim here is to compare the dynamics and limits of institutional constructions of regional integration in Africa, in relation to external connections, with the development issues associated with the vitality of exchanges from below. We then question the regionalization processes associated with artifacts based on international corridors and the transboundary management of environmental resources.

9.2. Regional paper integrations?

The continent has a multitude of supranational institutions, but their achievements are modest: the relative effectiveness of inter-African intervention forces does not mask intra-continental political divergences; the free movement of people and goods de jure is rarely de facto; economies remain highly dependent on world markets, while intra-continental trade and foreign direct investment (FDI) within Africa remain low.

9.2.1. *Slow progress in regional integration*

9.2.1.1. *The influence of the European model after independence*

Independence gave birth to some 40 states with borders inherited from colonial divisions. The federations of French West Africa (*Afrique occidentale française*, AOF) and French Equatorial Africa (*Afrique Équatoriale française*, AEF) broke up when the European project was taking shape. The construction of nation-states, the consolidation of national territories with inadequate infrastructure, and the imperatives of development filled the agenda of the young states, leaving little room for the construction of supranational economic and political organizations.

While the OAU (Organization of African Unity) and the two monetary unions in French-speaking Africa (UMOA and UDEAC[1]) that make up the Franc Zone date back to the 1960s (Pourtier 1993, 1995), it was not until the following decade that the regionalization process took shape, with the multiplication of economic communities, customs unions and watershed management organizations, and then, from the 1990s onwards, with international programs in energy and transport. Through the weight of its development partners and the major principles that guide it, the European Union undoubtedly influences regionalization in Africa, but it is carried out in a protean manner.

9.2.1.2. *The great bazaar of African regional institutions: overlaps, opportunistic strategies and weak implementation*

African regional institutions form a complex web (Figure 9.1). In West Africa, WAEMU (which replaced WAMU in 1994) forms a strong monetary union, backed by the Euro, but an economic community that is not very visible compared to ECOWAS[2], whose competencies are also exercised in the military (ECOMOG) and diplomatic domains, as well as in transportation (Bossard 2005) and infrastructure. In addition to these two entities with different contours, there are the management organizations of the large watersheds (OMVS, GRBDO, VBA, NBA[3]).

This complex landscape reveals leaderships (Nigeria for ECOWAS, South Africa for SADC), historical ties (WAEMU and CEMAC with France), the formalization of broad (CEN-SAD, COMESA) or narrow (IGAD, EAC, AMU[4]) international solidarities, the formulation of new projects (NEPAD[5]) or their repositioning (the AU replaced the OAU in 2002). In this continental geopolitics, regional institutions are both exposed to a form of competition among themselves (the AU, for example, recognizes only eight regional economic communities) and to the difficulty of reconciling divergent national policies and trajectories. Guinea and Mali are temporarily suspended from ECOWAS, which Mauritania left in 2000 to apply for observer status in 2017, as did Morocco and Tunisia. Within this community, free movement has been adopted by all 15 members, whereas in SADC, eight out of 15 members are still reluctant to do so (Commission économique pour l'Afrique 2017).

---

1. West African Monetary Union; Union of Central African States.

2. Economic Community of West African States.

3. Organisation de mise en valeur du Sénégal; Gambia River Basin Development Organization; Volta Basin Authority; Niger Basin Authority.

4. Intergovernmental Authority on Development; East African Community; Arab Maghreb Union.

5. New Partnership for Africa's Development, under the umbrella of the African Union.

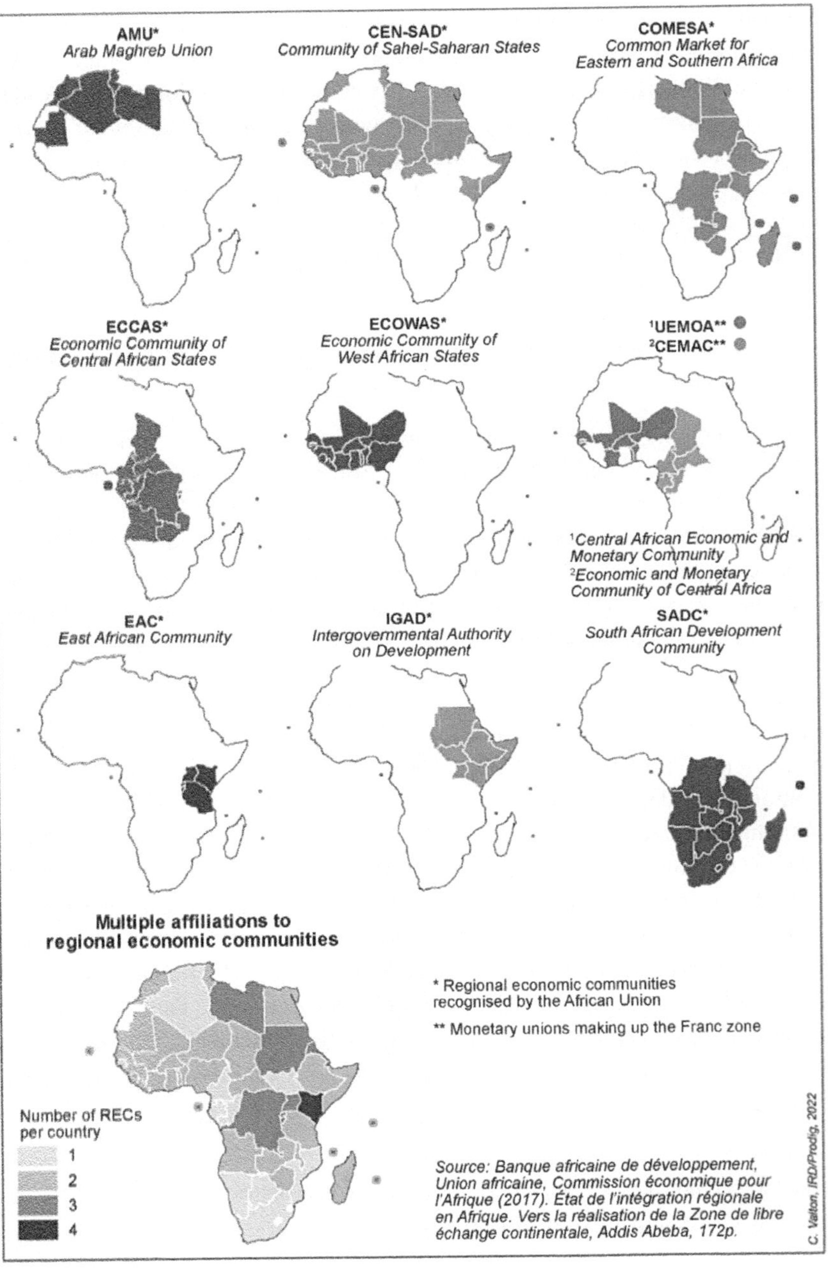

Figure 9.1. *The main regional institutions in Africa. For a color version of this figure, see www.iste.co.uk/beckouche/regionalization.zip*

9.2.1.3. *Recently relaunched regionalizations*

Since the 2010s, regionalization has found new sources of strength. The internationalization of several large African companies in the banking (Moroccan Attijariwafa), telecommunications (MTN), money transfer (Wari) and cement (Dangote) sectors is helping to create an African-centric economic fabric and accelerate the harmonization of business laws. Furthermore, the operationalization of the continental free trade area (FTAA) in 2021 is undoubtedly a turning point in the regionalization of the continent, by proposing to include and go beyond existing frameworks, extending the areas of international cooperation (to mining and health in particular) and laying the foundations of a pan-African common market (Banque africaine de développement 2017). Above all, having regained investment capacity thanks to a more favorable macroeconomic context and having forged new financial partnerships (with China, in particular), the states have launched vast infrastructure construction programs with an international vocation. The PIDA (Program for Infrastructure Development in Africa) supported by NEPAD is supporting 51 projects worth $360 billion in its first phase (2010–2020), half of which are either being financed or are underway. More than 73 other projects in a second phase are expected to be initiated by 2030 (Union Africaine 2020).

9.2.2. *Africa's regions as seen from the outside*

9.2.2.1. *Inherited blocks*

Regional representations of Africa have ancient roots. Following the Arab geographers of the Middle Ages, colonial representations of space distinguished between North Africa and sub-Saharan Africa, White Africa and Black Africa (Lacoste 1992), based on knowledge nourished by separate academic traditions, despite the alternative proposals of Diop (1954). The statistics of the United Nations still frequently repeat this dichotomy, distinguishing, for example, sub-Saharan Africa from the Arab States, which include, along with others, those of North Africa (PNUD 2020). Several recent atlases have nevertheless chosen to approach Africa as a continental-scale object (Losch et al. 2016; Magrin et al. 2018), recognizing the past, present and future importance of the relationship between its northern part and the rest of the African continent. The position of the Sahara as a hinge or border has fueled a literary production in France, in response to the question formulated in 2018 for the National Education competitions (Andrieu 2017; Choplin et al. 2017; Dumortier 2017).

Similarly, many contemporary institutional links are inherited: the southern African countries that emerged from the British Empire are members of the South African Development Community (SADC), one of the most dynamic African organizations, associated with the birth of the new South Africa after apartheid (1994). They are also members of the Commonwealth, which we can imagine will be more particularly invested in by the United Kingdom post-Brexit, as an instance of soft power of the Global Britain project. Similarly, the two currency blocs of the franc zone that emerged from AOF and AEF, WAEMU and CEMAC, were an essential part of the Organization of African, Caribbean and Pacific States (OEACP) created in 1975 to structure the trade partnership with Europe.

9.2.2.2. *A diversification of partnerships*

The new global geopolitics is making its mark on African regionalization. China has gradually integrated many African countries into its global plan for new maritime silk routes, starting in East Africa (Djibouti, Kenya, Ethiopia, Egypt), before extending this partnership since 2017 to North Africa (Tunisia, Morocco) and then West Africa (Senegal) (Pairault 2020).

France is using its influence through regionalized mechanisms that it is trying to Europeanize and Africanize. The G5 Sahel, an alliance formed in 2014 of five Sahelian countries fighting against armed Islamist groups, supported by Paris, is an example in terms of security. France also supported the evolution of the WAEMU towards a new currency, the ECO, in 2020, not without difficulty. Finally, voices are being raised in Europe in favor of an Africa-Mediterranean-Europe "vertical of the future" (Guigou and Beckouche 2017). The multidimensional integration of Europe with its Mediterranean and African peripheries is therefore thought to be the best way to channel, for shared benefit, the energy of a continent undergoing a demographic and urban surge – whose population will double by 2050 to reach 2.5 billion inhabitants, increasingly urban. Euro-African integration is both a remedy for African crises and an antidote to the decline of a demographically anemic Europe.

9.3. African trade from below

Data on African trade reveal a continent with little economic integration: intraregional trade accounts for less than 10% of ECOWAS trade and barely more than 20% for the EAC. But these figures do not take into account the multitude of informal, locally intense, and sometimes illicit trade, on which regional integration is based from below.

9.3.1. *Historical relational areas that replay in the present*

While regionalization "from above" generates new territorial configurations, integration "from below" relies on trade networks that reactivate relational systems that were dormant for a time. The colonial economy provoked a territorial inversion leading to the affirmation of coastal centralities and the marginalization of continental spaces and poles of exchange. "Upward" flows (of imported products from ports to inland cities) responding to "downward" flows (exported agricultural and mining products) have long dominated trade at the national and sub-regional levels, between landlocked and coastal countries. Since the 1990s, this matrix has become more complex under the combined effect of improved transport networks, economic diversification and the growth in African demand for consumer goods due to demographic and urban growth.

In West Africa, the diversification of urban–rural relations (Cottyn et al. 2013; Ninot and Sakho 2020) has been accompanied by a revitalization of trade circuits and hubs within the Sahelo-Sudanese continental spaces. Between Accra and Ouagadougou, the intensification of trade in agricultural products has thus strengthened the crossroad positions of the cities of Bobo-Dioulasso, Tamalé, Bittou or Bawku (Poujol 2017). On another scale, new trade relations (between Morocco and Senegal, for example), and also migratory flows and illicit trafficking (arms, drugs), have reactivated the old trans-Saharan routes and their crossroad cities (Marei and Ninot 2017). Around Lake Chad, the circulation of goods (livestock, agricultural products, manufactured goods), the driving force behind the regional system (Rangé 2016; Magrin and Pérouse de Montclos 2018), creates a complex system of relationships with both local and distant ramifications (from Tripoli to Douala and from Niamey to Port Sudan), taking advantage of an interface position between spaces that are differentiated from an agro-ecological, demographic, monetary and customs point of view. In East Africa, the city of Mbeya in Tanzania also takes advantage of its location at the contact of several borders (Malawi, DRC, Zambia, Mozambique) to articulate international exchanges, the supply of cities and the countryside with fish, cereals, fruits, vegetables and products imported from Dar es Salaam (Racaud and Mainet 2015). A geography is thus emerging in which the places of extraverted globalization (ports, metropolises) and those of regionalization centered on the continent are increasingly articulated (Giraut 2007).

9.3.2. *The ambiguous potential of national peripheries*

At the turn of the 1990s, the intensification of cross-border trade gradually structured vast territories located on the margins of national territories. They constitute "national peripheries", according to Igue (1995), whose existence and functioning seem to contradict both national constructions and institutional regional integration policies (Egg and Herrera 1998). Indeed, regional integration, based on a largely informal economy, is taking place around the borders of Nigeria, or in areas that replay old political and cultural identities, such as the Maradi–Katsina–Kano axis and the Sikasso–Korogho–Bobo-Dioulasso triangle (Enda Diapol 2007). In these densely populated areas crossed by major roads, where urban networks are relayed by numerous weekly rural markets, powerful economic actors (Walther 2006) linked by different identities (language, religion, ethnicity) and benefiting from the support of both local and national administrations, organize commercial exchanges whose particularity is to mix legal and illicit, long and short distances, imported and local products (Arditi et al. 1990).

Because they reveal a dissociation between national territories and trade areas, making more use of the opportunities offered by borders (monetary and fiscal differentials) than of complementarities between national economies that are not very complementary, the national peripheries are presented as alternative, partially autonomous spaces of integration that are the bearers of local development dynamics (monetary circulation, jobs). For all that, they are not the products of a possible anomie of states or of a supposed absurdity of borders (Bennafla 1999). On the contrary, the border is a resource and the state is an actor. The national peripheries benefit from the state's infrastructures and services, customs taxation benefits from trade, and security is a crucial issue for both the state and border users.

Faced with the dynamism of cross-border territories, states and regional institutions are torn. The temptation to try to formalize the informal sector remains strong (Karkaere et al. 2021), as does the temptation to capture customs revenues and secure porous borders where armed groups circulate (borders of Nigeria, Somalia and the DRC). However, total control is detrimental to the dynamism of the supply chains on which entire populations depend. For example, the closure of several borders during the Covid-19 pandemic undoubtedly limited the circulation of the virus, but it also had the effect of eliminating many formal and informal jobs, increasing the price of consumer goods and increasing food insecurity in many regions of sub-Saharan Africa (CEA 2020).

9.3.3. *The other side of the border*

Cross-border dynamism has its dark side. The trafficking that animates many African borders provides a living for thousands of border entrepreneurs, while at the same time undermining entire sections of the modern economy that form the backbone of development. The Compagnie sucrière du Sénégal (CSS) has thus developed since the 1970s under the customs protection of the Senegalese state in the classic framework of an import substitution agro-industry. The largest private employer in the country (approximately 8000 jobs in 2021) and in the medium-sized city of Richard Toll (60,000 inhabitants), it operates at the interface between the delta and the middle valley of the Senegal River, but lives under the constant threat of fraudulent sugar imports at world market prices, from the border areas of neighboring Mauritania or the Gambia, or through the port of Dakar. Similarly, in Cameroon, in 2011, the illegal export of 30,000 tons of cotton (20% of the national harvest) to Nigeria – where the world price had jumped – by thousands of motorcycle cabs brought Sodecoton, the jewel in the crown of the economy in the north of the country, to its knees, and underpinned its development model (supervision of producers, maintenance of tracks, food security, etc.) (Seignobos 2012).

Criminal activity also thrives on poorly controlled borders, weakening states while contributing to the existence of cross-border systems. Since the early 2000s, West Africa has become a hub of growing international cocaine trafficking between South America and Europe. Guinea Bissau, in particular, has served as a gateway and Mali as a relay. Drug money has increased the fragility of states (ONUDC 2008). However, cross-border criminal activity has a long history. In North Central Africa, highway robbers (zarguina), whose activities are part of an ancient tradition of rural banditry, used the borders to protect themselves from pursuit by regular armies. Their activities intensified with the crisis of the states in the 1980s and 1990s (Issa 2010). In crisis zones, the intensity of cross-border movements of armed men contributes to a regionalization through violence (Chauvin and Magrin 2020). From the Great Lakes region of Africa to the Sudan-Chad-CAR area, the profession of arms (Debos 2013) has become a professional horizon in states affected by a succession of conflicts (Pourtier 2003; Debos 2008). In the Sahel, the ferment of radical Islam is added to cross-border irredentism and criminal trafficking, to the point of constituting a regionalization factor. The geopolitical Sahel of the media, imported from American representations, no longer corresponds to the biogeographical representation of geographers (Alexandre 2020). It designates a vast Sahelo-Saharan strip that has the weakness of state control and the threat of armed Islamist groups on any globalized presence (and investment) in common. An anti-regionalization, so to speak.

9.4. Artifact-centered regionalizations

In parallel with institutional integration, multiple forms of international cooperation have been developed in key sectors (transport, the environment), enabling states to respond to, express, and even resolve possible disputes, and to give tangible form to regionalization.

9.4.1. *International corridors: a tool for extraversion or support for regional markets?*

9.4.1.1. *Persistent isolation*

The continent has 16 landlocked countries and suffers from the lowest road density in the world (less than 10 km/100 km^2 vs. 20 km in Asia and Latin America). Opening up the region remains a major challenge for states and regional institutions alike. Infrastructure networks have expanded everywhere since the 1960s (Lombard and Ninot 2010), but deficits persist (Foster and Briceno-Garmendia 2010). First, because the legacy of colonization, in the form of incomplete networks linking ports to productive hinterlands (Debrie 2007), was difficult to overcome. Second, because the planning policies of the young states were directed more towards consolidating national territories than towards international infrastructure. Similarly, bilateral agreements, which are fragile and fluctuating (such as those between Togo and its northern neighbors), have long prevailed over sub-regional agreements, which are superior on paper but less enforced on the ground. Above all, due to scarce and often poor infrastructure, as well as numerous illicit taxes (Teravaninthorn and Raballand 2009; Banque Africaine de développement 2015), international transport prices in sub-Saharan Africa are the highest in the world and make up to 70% of consumer prices for some imported products (Steck 2015).

9.4.1.2. *New generation corridors*

Since the 2000s, the corridor has become an indispensable instrument of regional integration and a lever for development at different levels, from the port metropolis to the sub-region, via the continental hub. The corridors have the advantage of bringing together public and private interests around colossal investments. The new port facilities have been financed by large international groups in exchange for operating concessions (Bolloré, Dubai Ports World in particular). However, the interests of economic operators are not always aligned with development issues. The corridors themselves carry a fundamental ambiguity, being more instruments for the extraversion of economies than supports for intra-African trade. The Ivorian railway operated by the Sitarail company (Bolloré Group) has thus become a very efficient tool for international transit traffic between the port of Abidjan and Ouagadougou, at the cost of abandoning local services, which are nonetheless essential to national and regional economies (Dagnogo et al. 2012). The corridor model, a product of neoliberal logic, radically opposes the imperatives of the international economy and those of territorial development (Lombard et al. 2014). The impact of the large investments it implies also threatens the environment, as in Kenya (Fourault-Cauët and Steck 2019).

The context is nonetheless favorable to major works: the visions provided by the African regional impetus (Nepad, FTAA) are being met by African (AfDB) and international (World Bank) institutions, old (EU) and new (China) financial partners, and the states themselves (ICA 2018). Huge projects are emerging, such as the Lapsset corridor (Lamu port, South Sudan, Ethiopia), which by 2030 will form a multimodal axis (rail, road, air, pipeline and fiber) opening up new prospects for this part of Africa. In this continent under construction (Magrin et al. 2018), regional networks are at different stages of maturity: dense and well connected in SADC, under consolidation in ECOWAS and East Africa, and still loose and disjointed in Central Africa.

9.4.2. *Environmental integration?*

9.4.2.1. *Hydrosystems: between conflicts and compromises*

Since the 1960s, the management of transboundary hydrosystems has provided a concrete horizon for regional cooperation in Africa (Raison and Magrin 2009). For example, the OMVS, involving Mali, Mauritania and Senegal, was able to channel significant international aid after the Sahelian droughts of the 1970s and 1980s to build the Diama anti-salt dam (1986) in the lower delta and the Manantali reservoir (1988) on the Bafing River, allowing for the development of irrigated agriculture and the production of hydroelectricity (2003), for the benefit of the three partner countries. This exemplary regional cooperation did not, however, prevent the deterioration of relations between neighbors in 1989, when the interweaving of local tensions and national crises almost led to a war between Senegal and Mauritania (Seck et al. 2009).

Elsewhere, hydraulic cooperation illustrates geopolitical situations more than the need for joint management of the resource. In southern Africa, hydraulic cooperation, technically very accomplished, serves above all the interests of the South African power (Blanchon 2009). Around Lake Chad, the creation of the Lake Chad Basin Commission (LCBC) in 1964 has hardly led to any concrete progress. A utopian megaproject to transfer water from the Congo Basin, intended to save a Lake Chad presented as being under threat of drying up, constitutes the fragile cement of a regional organization torn between member countries (Cameroon, Niger, Nigeria, CAR, Chad, Libya) beset by multiple difficulties, and for which water management is not the priority (Magrin and Lemoalle 2019). On the Nile, weak cooperation risks leading to conflict. The Great Ethiopian Renaissance Dam, built unilaterally on the

Blue Nile (2011–2019) by an Ethiopia that denounces the colonial rules (1929, 1959) for sharing the river's waters, is dangerously poisoning its relations with downstream countries, particularly Egypt, for which the great river is vital (Blanchon 2019; Melesse et al. 2021).

9.4.2.2. *The complex dialogues of conservation*

Conservation forms another sensitive issue of regional cooperation. In 2016, 18 Transfrontier Conservation Areas were recognized by the SADC. After apartheid, these peace parks aim for integrated management of large wildlife and tourist traffic. The infrastructures and associated jobs are supposed to promote local economic development and the adherence of the populations to the preservation of biodiversity. Yet the promoters and main beneficiaries of this approach are South Africans (NGOs, parks and tourism companies) (Guyot 2004; Belaidi 2016).

The Great Green Wall is another environmental regionalization mirage (Mugelé 2018). This project, first promoted by Senegalese President Wade in the mid-2000s, involving a dozen countries, aimed to form a 15-km wide tree belt from Senegal to Djibouti, over 7,000 km, in order to combat desertification in the Sahel. Its actual implementation is modest. At best, the project has been able to channel some funding for reforestation operations or support for market gardening, as in the Senegalese Ferlo. In this pastoral territory, however, it was conceived without integrating the needs of the herders. This project thus appears to be a "glocal" object, resilient on the international climate change scene, but less likely to strengthen cooperation between countries sharing the same challenges than to stir up rivalries to capture adaptation rents (Magrin and Mugelé 2020).

9.5. Conclusion

Francophone geography has been a pioneer in understanding the tension between the logics of institutional integration and the dynamics of integration "from below".

Since the beginning of the 21st century, continental dynamics (demographic and urban dividends) and global dynamics (growing Chinese influence and the imperative of transition) have been drawing new horizons and suggesting new prospects to African States. Can the new regional momentum create the right framework to transform these prospects into development dynamics? With a prospective population of 2.5 billion in 2050, the FTAA presents itself as a promising market and employment area, provided that major challenges are quickly addressed: energy, infrastructure, education and essential services and security. While some of these challenges are multilevel competencies, from the infra to the supranational level, they depend mainly on the power of the states. The economic (rentier systems) and political fragility of states is not diminishing, especially since the decentralization process underway in most countries and the fragmentation underway in some of them (Ethiopia, Somalia, Sahelian states) are leading to a narrowing of ambitions and powers at the local level.

Continental specificities suggest that countries and regional economic communities will follow specific trajectories rather than known models; however, it is difficult to imagine that RECs alone can drive economic development and build the conditions for it to occur. On the contrary, the acceleration of continental regional momentum at the institutional level and the local amplification of cross-border trade may fuel resistance both at the level of the states themselves and at the level of sub-regional organizations seeking legitimacy, sovereignty and control. One of the main challenges therefore remains that of coordination between the different levels of political decision-making and economic dynamism.

From a scientific point of view, the question is to know whether the new situation can be accommodated by an adjustment of the classic questions (e.g. on the tension between institutional integration and dynamics from below) or whether it justifies new epistemologies.

9.6. References

Andrieu, J. (ed.) (2017). *L'Afrique : du Sahel et du Sahara à la Méditerranée*. Ellipses, Paris.

Bach, D. (1998). *Régionalisation, mondialisation et fragmentation en Afrique sub-saharienne*. Karthala, Pairs.

Banque africaine de développement (2015). Problématique de la facilitation du transport en Afrique de l'Ouest et Plan d'actions. Report, Abidjan.

Banque africaine de développement, Union africaine, Commission économique pour l'Afrique (2017). Etat de l'intégration régionale en Afrique. Vers la réalisation de la zone de libre échange continentale. Report, Addis Abeba.

Bayart, J.F. (2006). *L'État en Afrique. La politique du ventre*, 1st edition. Librairie Erthème Fayard, Paris.

Belaidi, N. (2016). La paix par la coopération environnementale, un nouveau modèle de conservation ? Etude des aires transfrontalières en Afrique australe. *Afrique contemporaine*, 257, 129–143.

Bennafla, K. (1999). La fin des territoires nationaux ? Etat et commerce frontalier en Afrique centrale. *Politique Africaine*, 73(1), 25–49.

Besnoist, J.R. (1979). *La balkanisation de l'Afrique occidentale française*. Nouvelles éditions africaines, Dakar.

Blanchon, D. (2009). L'Orange ou l'inégal partage des eaux australes. In *Des fleuves entre conflits et compromis. Essais d'hydropolitique africaine*, Raison, J.P. and Magrin, G. (eds). Karthala, Paris.

Blanchon, D. (ed.) (2019). *Géopolitique de l'eau : entre conflits et coopérations*. Le Cavalier Bleu, Paris.

Bossard, D. (2005). Atlas régional des transports et des télécommunications dans la CEDEAO. Report, CEDEAO, CSAO, Abuja, Issy-les-Moulineaux.

Choplin, A., Pliez, O., Mareï, N. (eds) (2017). *L'Afrique : du Sahel et du Sahara à la Méditerranée*. Atlande, Neuilly.

Commission économique pour l'Afrique (2020). Faciliter le commerce transfrontalier par la coordination de la riposte africaine au Covid-19. Report, CEA, Addis-Abeba.

Cottyn, I., Schapendonk, J., Van Lindert, O. (2013). Mobility in sub-Saharan Africa: Patterns, processes, and policies. Report, RurbanAfrica [Online]. Available at: https://rurbanafrica.ku.dk/publications/briefings/2013/RurbanAfrica-Briefing-Paper2.pdf.

Dagnogo, F., Ninot, O., Chaléard, J.-L. (2012). Le chemin de fer Abidjan-Niger : la vocation d'une infrastructure en question. *Echogéo*, 20 [Online]. Available at: https://journals.openedition.org/echogeo/13131.

Debos, M. (2008). Darfour, Tchad, République centrafricaine : le développement d'une crise régionale. In *Ramses 2008*, de Montbrial, T. and Moreau Defarges, P. (eds). Dunod, Paris.

Debos, M. (2013). *Le métier des armes au Tchad. Le gouvernement de l'entre-guerres*. Karthala, Paris.

Debrie, J. (2007). Géohistoire d'un tracé technique : frontières et réseaux en Afrique de l'Ouest continentale francophone. *Flux*, 70, 49–58, October/December.

Diop, C.A. (1954). *Nations nègres et cultures*. Présence africaine, Paris.

Dumortier, B. (ed.) (2017). *L'Afrique : du Sahel et du Sahara à la Méditerranée*. Armand Colin, Paris.

Egg, J. and Herrera, X. (ed.) (1998). *Échanges transfrontaliers et intégration régionale en Afrique subsaharienne*. Autrepart. Editions de l'Aube, Orstom.

Enda Diapol (2007). *Les dynamiques transfrontalières en Afrique de l'Ouest. Analyse des potentiels d'intégration de trois "pays-frontières" en Afrique de l'Ouest*. CRDI and Karthala, Paris.

Foster, V. and Briceno-Guarmendia, C. (2010). Infrastructures africaines. Une transformation impérative. Document, AFD, Banque mondiale.

Fourault-Cauët, V. and Steck J.-F. (2019). Corridor contre corridor : Nairobi. *EchoGéo*, 49 [Online]. Available at: http://journals.openedition.org/echogeo/18006.

Giraut, F. (2007). L'Afrique et ses territoires, un continent en mondialisation. In *La mondialisation, l'intégration des pays en développement*, Cadene, P. (ed.). SEDES, Paris.

Guyot, S. (2004). Derrière l'écotourisme, le politique : conservation et discrimination territoriale en Afrique du sud. *Revue Tiers Monde*, 178, 341–363. doi: 10.3917/rtm.178.0341.

Hélène, M. and Sylvain, R. (2016). Secondary towns in globalization: Lessons from East Africa. *Articulo, Journal of Urban Research*, 12 [Online]. Available at: http://journals.openedition.org/articulo/2880.

Igue, J. (1995). *Le territoire et l'État en Afrique*. Karthala, Paris.

Issa, S. (2010). *Les coupeurs de route. Histoire du banditisme rural et transfrontalier dans le bassin du lac Tchad*. Karthala, Paris.

Karkare, P., Byiers, B., Apiko, P., Kane, M. (2021). Pas une erreur, mais un système : le commerce informel transfrontalier en Afrique de l'ouest. Document, ECDPM and Enda-Cacid.

Lacoste, Y. (1992). Afriques noires, Afriques blanches. *Hérodote*, 65–66.

Lombard, J. and Ninot, O. (2010). Connecter et intégrer. Les territoires et les mutations des transports en Afrique. *Bulletin de l'association des géographes français*, 1, 69–86.

Lombard, J., Ninot, O., Steck, B. (2014). Corridors de transport en Afrique et intégration territoriale en questions. In *La régionalisation du monde. Construction territoriale et articulation global/local*, Alia, G. and Yann, R. (eds). IRMC, Karthala, Paris.

Magrin, G. (2013). *Voyage en Afrique rentière. Une lecture géographique des trajectoires du développement*. Publications de la Sorbonne, Paris.

Magrin, G. and Lemoalle, J. (2019). L'avenir du lac Tchad. Les échelles de l'incertitude. In *Le Tchad des lacs. Les zones humides sahéliennes au défi du changement global*, Raimond, C., Sylvestre, F., Zakinet, D., Moussa, A. (eds). IRD-Editions, Marseille.

Magrin, G. and Mugelé, R. (2020). La boucle de l'Anthropocène au Sahel : nature et sociétés face aux grands projets environnementaux (Grande Muraille Verte, Sauvegarde du lac Tchad). *Belgeo*, 3 [Online]. Available at: http://journals.openedition.org/belgeo/42872.

Magrin, G. and Pérouse de Monclos, M.-A. (2018). *Crise et développement. La région du lac Tchad à l'épreuve de Boko Haram*. AFD, Paris.

Magrin, G., Dubresson, A., Ninot, O. (2018). *Atlas de l'Afrique. Un continent émergent ?* Autrement, Paris.

Mareï, N. and Ninot, O. (2018). Entre Afrique du Nord et de l'Ouest, les relations transsahariennes à un moment charnière. *BAGF*, 95(2) [Online]. Available at: https://journals.openedition.org/bagf/3189.

Melesse, A.M., Abtew, W., Moges, S.A. (2021). *Nile and Grand Ethiopian Renaissance Dam. Past, Present and Future*. Springer, New York.

Mugelé, R. (2018). La Grande muraille verte au Sahel : entre ambitions globales et ancrage local. *Bulletin de l'association de géographes français*, 95(2) [Online]. Available at: http://journals.openedition.org/bagf/3084.

Ninot, O. and Sakho, P. (2021). Évolution des relations villes-campagnes en Afrique de l'Ouest : une lecture à partir des circulations des personnes et des biens. In *Crises des modèles ? Agricultures, recompositions territoriales et nouvelles relations villes-campagnes*, Berger, M., Chaléard, J.-L., Gana, A. (eds). Grafigéo-PRODIG, Paris.

Pairault, T. (2020). Examining the importance of the new silk roads for Africa and for global governance. In *The Belt and Road Initiative and Global Governance*, Carrai, M.A., Defraigne, J.-C., Wouters, J. (eds). Edward Elgar Publishing, Cheltenham.

PNUD (2020). Rapport sur le développement humain 2020. La prochaine frontière : le développement humain et l'Anthropocène. Report, PNUD, New York.

Poujol, G. (2017). Les circuits vivriers du corridor Ouagadougou – Accra : conditions d'un développement inclusif. PhD Thesis, Université de Montpellier 3.

Pourtier, R. (1993). Atlas de l'Udeac. Ministère de la coopération, Paris.

Pourtier, R. (1995). Atlas de la zone franc en Afrique subsaharienne. Ministère de la coopération and La Documentation française, Paris.

Pourtier, R. (2003). L'Afrique centrale dans la tourmente : les enjeux de la guerre et de la paix au Congo et alentour. *Hérodote*, 111, 11–39. doi: 10.3917/her.111.0011.

Rangé, C. (2016). Multi-usage des ressources et mobilités : l'intensification dans une zone humide sahélienne. Le lac Tchad vu par sa fenêtre camerounaise. PhD Thesis, AgroParisTech.

Sautter, G. (1958). Le chemin de fer Bangui-Tchad dans son contexte économique régional. Société civile d'études du chemin de fer de Bangui au Tchad, Bangui.

Seignobos, C. (2012). La moto chinoise. Une révolution urbaine et rurale. In *Les échanges et la communication dans le bassin du lac Tchad : actes du colloque de Naples du réseau Méga-Tchad*, Baldi, S. and Magrin, G. (eds). Universita degli studi di Napoli "L'Orientale", Naples.

Steck, B. (2015). Introduction à l'Afrique des ports et de corridors : comment formuler l'interaction entre logistique et développement. *Cahiers de géographie du Québec*, 59(168), 447–467.

Teravaninthorn, S. and Raballand, G. (2009). *Transport Prices and Costs in Africa: A Review of the International Corridors*. The World Bank, Washington DC [Online]. Available at: https://openknowledge.worldbank.org/bitstream/handle/10986/6610/461810PUB0Box3101OFFICIAL0USE0ONLY1.pdf?sequence=1&isAllowed=y.

Union Africaine (2020). Rapport d'avancement du PIDA 2019–2020. Report, Union Africaine.

Walther, O. (2006). Affaires de patrons, usages de la frontière. Géographie des réseaux marchands entre Niger, Bénin et Nigeria. PhD Thesis, Université de Rouen, Université de Lausanne.

10

Europe, a Geographical Puzzle (A)

Etienne TOUREILLE[1], Antoine LAPORTE[2] and Claude GRASLAND[3]

[1] *UMR 6266 IDEES, Université de Rouen – Normandie, France*

[2] *UMR 5600 EVS, École Normale Supérieure de Lyon, France*

[3] *UMR 8504 Géographie-cités, Université de Paris, France*

10.1. Introduction

The polysemy of the word "*Europe*" covers a diversity of geographical objects with different semantic contents and geometries. It can sometimes be assimilated to the European Union (EU), and can also designate a hypothetical continent (Europe "from the Atlantic to the Urals") and a less clearly defined object (a "large region", "macro-region" or "supranational region", a "territory"), whose limits would have to be specified by the geographer. Few regions of the world see so many debates flourish.

This is a difficult task, as collective representations differ according to the training, experiences or social position of individuals. Like any word designating a regional group, it is both commonly used and abstract. It can refer to a social as well as a political construct. Europe designates both a space, an area of reference that allows us to orient ourselves on the emerging lands of the globe (the western peninsula of the Eurasian ensemble bordered by the Atlantic, the Arctic Ocean and the Pacific Ocean), and the territories of supra-state associations such as the EU.

Regional analysis in geography is only a minority component of the field of European studies, in which other claimed currents of geography (theoretical and

Regionalization of the World,
coordinated by Pierre BECKOUCHE and Yann RICHARD.

quantitative geography, economic geography, social geography, geohistory) and other disciplines (international relations, political science, economics, law) are joined. Its specificity lies in the approach of putting the correspondence between the notion of "Europe" and the space (its limits, its content) that it designates at the heart of its questioning.

This chapter adopts a nominalist posture by considering "Europe" as a notion that exists only in the words and language conventions used to describe it, a region that geographic analysis aims to define.

10.2. The various definitions of Europe

10.2.1. *Europe captured by the EU: the institutional tropism*

In everyday language, the word "Europe" is often used as a synonym for EU (Figure 10.1). This equivalence is found in media (see section 4.2), political and academic discourse, and leads to a pre-eminence of the EU in European studies, which English-speaking scholars point to through the pun "*EUrope*".

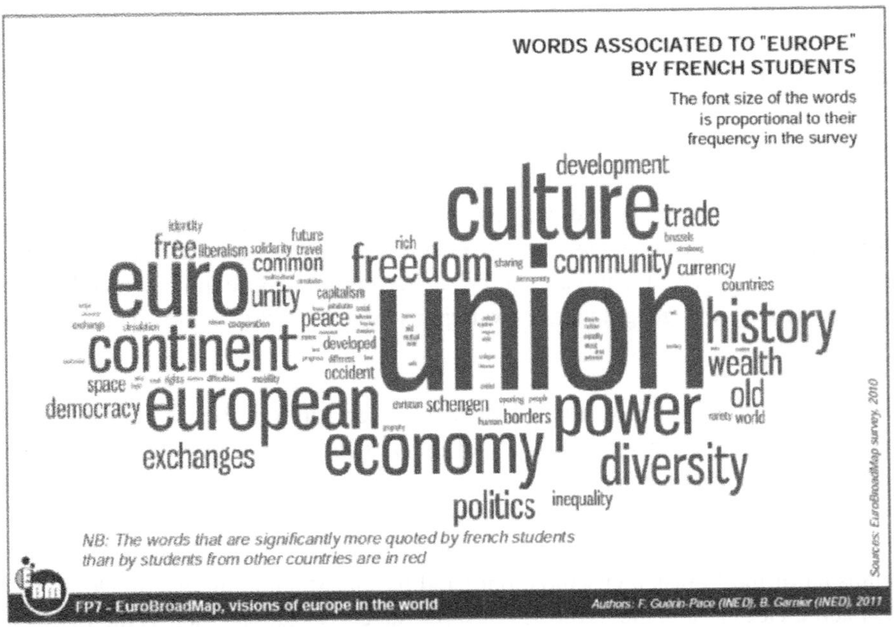

Figure 10.1. *Terms associated with the word "Europe" in the Eurobroadmap survey of French undergraduate students (source: Grasland (2011)). For a color version of this figure, see www.iste.co.uk/beckouche/regionalization.zip*

10.2.1.1. *A "EUropeanized" research object*

Research in geography is no exception: the analysis of the role of the EU and its policies in regional integration has been central to the work of specialists in this part of the world for half a century.

At the beginning of the 21st century, the EU emerged as a funder of scientific research through two different mechanisms. First, the EU developed its own policy for funding scientific research through the European Research Council and the launch of "Europe Horizon 2020" projects by the European Commission. These funding windows lead thematic geographers to consider themes relevant to the European scale (metropolization, cross-border issues, international migration, spatial interaction, regional integration). The other process is linked to the development of research in geography applied to European territorial planning. With the adoption in 2007 of the "Charter on Sustainable European Cities" and the European "Territorial Agenda" by the European Council of Ministers, the EU has adopted a spatial planning policy. These decisions lead to the political legitimization of the production of expert reports on macro-regional planning issues of the European territory, which were initiated in 1997 with the adoption of the European Spatial Development Perspective (ESDP), and reinforced in 1999 by the establishment of a European Spatial Planning Observatory Network (ESPON).

As was historically the case with the DATAR's action for the French territory, this applied research within the framework of ESPON will become a space for scholarly production at the interface between political orders and the academic sphere. Typically French, this planned approach to territorial cohesion has been ironically described as "old French wine in new bottles" (Faludi 2004). The existence of funding in this field will push researchers who do not always identify themselves a priori as specialists in European studies to develop a specifically localized production in this territorial framework. This space of expression has sometimes been built independently from the classic national university research frameworks (CNRS research units, universities), while the European scope of these projects has led to cooperation between researchers from different institutions and countries, often also across disciplinary boundaries.

The "EUropeanization" of academic research in geography, and particularly in regional and political geography, has thus contributed to bringing the EU to the forefront as a privileged territory of analysis[1] – without, however, leading to the emergence of a clearly structured field of research in francophone research, as many researchers remain within their original thematic affiliations (migration studies, urban studies, environmental geography, for example).

10.2.1.2. *The EU as a framework for analysis and as a player in regional integration*

The EU can thus first of all play the role of a simple analytical framework defined a priori for the study of a targeted theme: for example, poverty in Europe (Boulineau and Bonerandi 2014), shrinking cities and regions (Baron et al. 2010; Bailleul et al. 2019), planning issues (Jean and Baudelle 2009). It can also become an object of study in itself through the analysis of its spatial translations, such as regional integration: territorial convergence (decrease in equalities between territories, increase in common characteristics), increase in interdependence between territories through cross-border mobilities, student mobilities, monetary circulations (Le Texier 2014), the diffusion of legal or administrative norms in the territories of member, candidate or neighboring countries. In the latter case, we speak of formal integration, sometimes referred to as "Europeanization" (Jessop 2002; Radelli 2004; Bafoil 2006, 2008), which reinforces the ambiguity between what belongs to the EU and more generally to "Europe". For example, the creation of regions in Turkey to comply with the requirements of the EU's regional policy, a country where the regional level for statistical and administrative management did not exist (Montabone 2013). In these meanings, the idea of Europeanization takes on a normative dimension, becoming the framework and operating mode of a European model (Faludi 2007). Its analysis thus becomes a means of evaluating the conformity of territories to this model, an operation that is therefore far from neutral.

Others are more interested in formal integration: how transnational partnerships are woven to bring about the emergence of regional groupings, through the study of the circulation of models in territorial marketing, the dissemination of planning and local development engineering practices, and the way in which these different practices participate in the construction of imaginaries of the EU and its territories (Escach 2015; Laménie 2019; Briot 2021; Gourdon 2021).

This work is therefore anchored at different scales. The EU can be studied on a macroscopic scale by carrying out quantitative analyses on the scale of a set of NUTS (Nomenclature of Territorial Units for Statistics) of the EU, in order to reflect

1. Even though in the case of ESPON, the study area is extended to non-member countries such as Switzerland, Norway, Liechtenstein or Iceland, but not Turkey, although it is a candidate.

on the problems of territorial cohesion (Faludi 2004; Grasland and Hamez 2005; Bourdin 2020). We can also work on a more local scale by reflecting on the way in which this new transnational level is constructed, from a department of a municipality, a regional council, a consultancy firm or an association. Studies of the EU are not limited to the official limits of the EU: a quantitatively limited but politically important part of the work focuses on relations with its regional environment, notably through the study of the European Neighborhood Policy (ENP – Richard 2008; Beckouche and Richard 2008, 2013; Beckouche et al. 2016), "North/North" cooperation frameworks and Euro-Mediterranean issues (Beckouche 2008). The aim here is to take an interest in the various configurations of partnerships that contribute to shaping a "variable geometry" EU-European territory. For the past 15 years, for example, the MIGREUROP collective has been studying the regional ensemble constructed by the management of the Schengen area border, through the analysis of the externalization of this border (Migreurop 2017). The space of surveillance and coercion highlighted by maps of detention camps and bilateral partnerships with countries responsible for outsourcing border management (Libya, Ukraine, Mali, for example), lead to regional forms that cover a wide "Eurasian" fringe.

In the end, a multitude of works question the limits of the space influenced by EU policies, making them the dominant object of study in European regional and political geography studies. However, even though the EU is a key object of study for discussing the emergence of a supranational regional level, the EU alone is not "Europe".

10.2.2. *Europe-continent: a cumbersome ethnocentrism*

The other object that occupies a predominant place in the representations of "Europe" is that of the European continent, particularly in popular representations and political discourse. Continentality, whether explicit or not, constitutes a recurrent subtext in the way of considering the definition of a hypothetical European region.

10.2.2.1. *Europe-continent: an obvious social representation…*

Since the mid-2000s, several studies conducted by specialists in world representations have highlighted the salience of the notion of continent in the representations that the inhabitants of EU member countries (or candidate countries) have of themselves and of the territory to which they belong (Didelon 2010; Didelon et al. 2011, 2014, 2018; Toureille 2017; Figure 10.2).

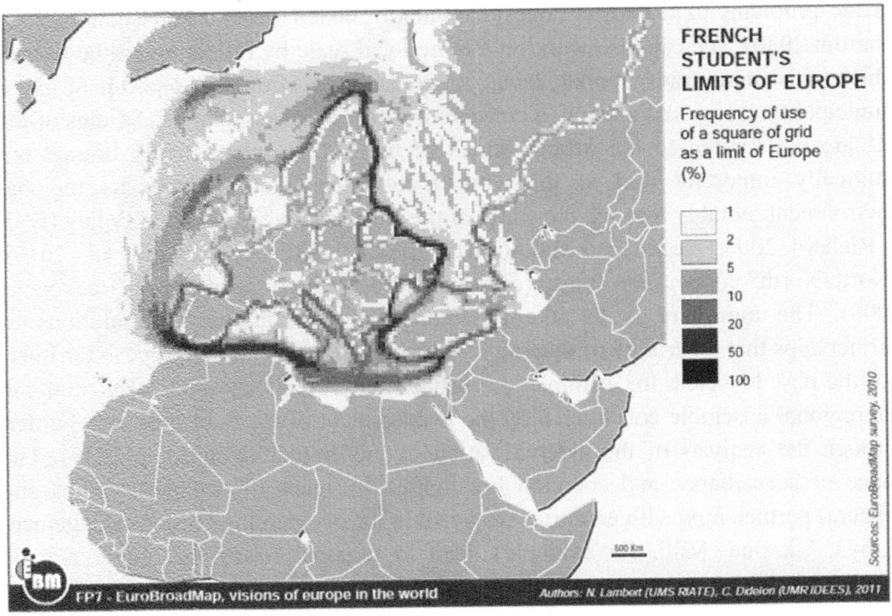

Figure 10.2. *The limits of "Europe" in the Eurobroadmap survey of French undergraduate students (source: Grasland (2011)). For a color version of this figure, see www.iste.co.uk/beckouche/regionalization.zip*

The notion of the continent also occupies a prominent place in the political imagination, as underlined by the 2005 referendum campaign on the adoption of the *Treaty Establishing a European Constitution* in France. During this rare moment when the political news of the EU was put on the front stage, the campaign led to a lively debate in France about the limits of Europe. In the wake of the EU's biggest-ever enlargement (10 new members in 2004), the forthcoming integration of Turkey, which became an official candidate for EU membership (Helsinki Summit, December 1999), became the campaign theme of the conservative right. This debate led a number of leading politicians to propose geographies of "Europe" on which this political construction should be based (Cautrès and Monceau 2011):

> Turkey is a country close to Europe, an important country, which has a real elite, but it is not a European country. Its capital is not in Europe, it has 95% of its population outside Europe, it is not a European country (V. Giscard d'Estaing, interview for *Le Monde*, November 9, 2002).

The argument that 95% of Turkish territory was in "Europe", as its capital is located in "Asia Minor" and not in "Europe", seemed to refer to geographical evidence: the model of a continental "Europe", "from the Atlantic to the Urals", from Scandinavia to the Mediterranean. The said treaty, which was never finally ratified, left no room for doubt: among the 13 occurrences of the word "Europe" in the text, one of them explicitly refers to the term continent, alongside the no less problematic term civilization:

> Aware that *Europe is a continent* [emphasis added] which is the bearer of civilization; that its inhabitants, who have come in successive waves since the earliest times, have progressively developed the values on which humanism is founded: the Equality of Beings, liberty, and respect for reason (Preamble of Part I of the *Treaty Establishing a European Constitution*, 2005).

10.2.2.2. *...but problematic as a category of analysis*

This continental vision of the European region nevertheless poses a scientific problem. Intervening in the political debate, the geographer P. Clerc qualifies it in an article in the *Le Monde* of November 18, 2002 as a "pre-text geography", i.e. an element of language used above all for political purposes, historically situated and unsuited to contemporary geography. He reminds us that the continents are the fruit of the encyclopedic and classificatory thought that ran from the end of the 18th century to the beginning of the 20th century. Borrowed from a naturalist logic, very much in vogue at the time, their delimitations were based on landmarks drawn from topography (closed seas, rivers, mountain ranges). They give the illusion of their fixity and their "natural" character, when some of them, like the Urals, the theoretical border between Europe and Asia, were conceived in the antechamber of chancelleries: the geographer who is credited with the invention of this, V.N. Tatishchev, responded to an order by Tsar Peter Ist to geographically manifest the entry of the Russian empire into the concert of nations of Europe. These continental visions, proceeding through space and cartography as they are consulted in atlases and textbooks, add to the semantic sedimentation of an already polysemous notion, drawing its roots from Greek and biblical cosmogonies and mythologies (Parker 1960; Ahnström 1993; Lewis and Wigens 1997; Grataloup 2009).

However, these continental visions are too small or too large to account for the influence of a whole range of current phenomena, such as energy supply, trade interdependencies (Figure 10.4), tourism and capital flows (see Figure 10.5). Finally, continental categories have the disadvantage of being Eurocentric objects: Europeans divided up the world at a time when they considered themselves to be at its center (Amin 1988; Grataloup 2009). Attempts to objectify them generally lead to failure or to contradictory results that do not conform to common sense representations, as shown by the potential and surface maps in Figure 10.3.

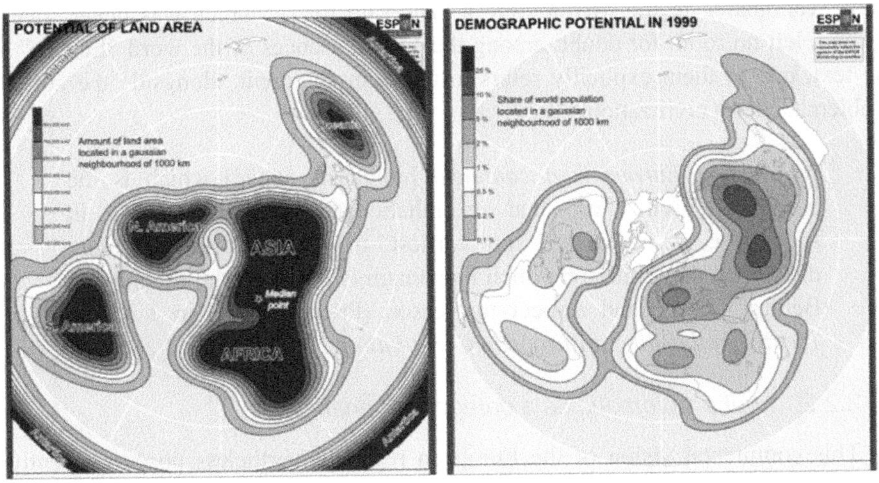

Figure 10.3. *Surface and population potential maps (source: Beckouche et al. (2008)). For a color version of this figure, see www.iste.co.uk/beckouche/regionalization.zip*

10.2.3. *The geographer's Europe: a whole to be (de)constructed, to be regionalized*

The study of "Europe" in regional geography must therefore accommodate these two cumbersome objects, which is not easy. This is evidenced by the recent *Dictionnaire de la régionalisation* (Mareï and Richard 2018), which does not include in its index any "Europe" entry, but entries for "European Union", "Eurasia" (a term that is also fraught with meaning in Turkey and the post-Soviet space), "Eurafrica", "Euromed" and "Europeanization". It includes, on the contrary, entries on "Africa" and "Americas". Others prefer to work on "the EU and its neighborhood" (Beckouche and Richard 2008), thus envisaging a whole constructed in interaction between an identifiable political core and various peripheries via relations of various natures (economic exchanges, political relations, dependence, domination). Still, others get around the difficulty by implementing strategies to avoid the words "EU" and "Europe":

> At the risk of sounding provocative, the solution that would have been preferred by the authors would have been to fully assume the variable or indecisive nature of the space covered by the analysis and to

replace the occurrences of the terms "Europe" or "European Union" with a symbol of absence of definition ("***") or with an arbitrarily chosen word such as "Dzom" (Didelon et al. 2008).

In all cases, "Europe" is the beginning or the end of something: sometimes "a small cape of Asia" (Valéry 1919), sometimes "a vast peninsula at the western end of the Eurasian block, as India is in the South" (Vandermotten and Dézert 2008, p. 1). The terminology used to designate the spatial object under study varies: "macro-region" (Gana and Richard 2014; Mareï and Richard 2020), "large region" (Demangeon 1932; Girault *in* Mareï and Richard 2018), "greater Europe" (Beckouche and Richard 2013), "European territory" (Didelon et al. 2011), "part of the world", etc. All of these qualifications do not explicitly refer to the notion of region and undoubtedly reveal a conceptual weakness in the definitions proposed by geography.

To escape the previous aporias, a scalar reasoning could pose the question in the following way. "Europe" is either *a part of the* world: the highest level of division below the world (the highest level of observation); or a set of supra-state territories: the first level above the states. Either perspective will obviously not lead to identical results. In all cases, what is at stake is the heuristic dimension of the object: its form, its limits, its content and its different components constitute the central element of the analysis. Following on from this point, the concept of region can constitute a milestone of formalism useful to analyze the object. We propose here to use the term "macro-region", which for some years now seems to be the consensus among geographers who propose to theorize these intermediate objects between the state and global levels (Mareï and Richard 2020).

10.3. Finding the limits of Europe

From this point of view, several perspectives exist to define plural or variable geometric "Europe", valid both for works concerning the EU alone and for those aiming at a larger whole. The analysis will focus mainly on the study of the external limits of Europe, leaving the analysis of its internal divisions aside.

10.3.1. *A plurality of approaches*

Some of the major currents that have contributed to this field since the early 2000s are presented here, without claiming to be exhaustive.

10.3.1.1. *The critique of a homogeneous cultural region: the sidestepping strategy of geohistory*

The perspective offered by geohistory is based on a temporal decentering, highlighting the historically contingent character of continental divisions and the different cultural constructions with which they are associated (civilizations, the West, etc. – Anhnström 1993; Lewis and Wigen 1997; Grataloup 2009). "Europe" does not have the same meaning depending on the country and time in which one is situated (Grataloup 2009; Brennetot and Rosenberg 2013). In the francophone context, this Braudelian-inspired geography is linked to the more general question of the study of the world-system, its division into regional subsets and globalization (Dollfus 2001; Grataloup 2007; Wallerstein 2009; Capdepuy 2018). Most of the time, it develops a critical perspective of the "grand narratives" that structure school teaching or political discourse.

10.3.1.2. *Functional Europe captured by flows and the complementarity of spaces*

Geographies of politics study Europe through the study of the stakes involved in technical devices: fluid transport networks, air networks, standards or legal devices. This is often the work of specialists in the geography of neighborhoods or of targeted themes (e.g. in the field of transport, see Beyer (2020)). In the case of the geography of energy, we will consider, for example, the political construction of a European space through the integration of its supply network (see Palle (2017) and Chapter 3 of this book). These works are most often based on the analysis of official sources (grey literature, institutional communication, interviews with actors) or media sources, sometimes using discourse analysis methods. They are more rarely based on quantitative analyses of the flows and networks concerned.

10.3.1.3. *Quantitative approaches: mathematical abstraction to construct homogeneous or functional regions*

Spatial analysis and data analysis, on the contrary, propose alternative delimitations based on sets of statistical indicators. The regional forms considered depend on the research object chosen: functional regions in the case of city systems, graphs representing common votes at United Nations assemblies (Beauguitte 2011), transport flows, homogeneous regions through socio-demographic, economic or human freedom measurement indicators (Beckouche et al. 2008; Didelon et al. 2008; Grasland and Van Hamme 2010; Toureille 2020).

From a theoretical point of view, these approaches have the advantage of not necessarily positing a priori the existence of a European region. They can in fact start from a matrix of world trade that will be partitioned inductively into integrated territories according to reproducible criteria. It may be that none of the territories discovered at the end of the analysis cover the EU, or even that the EU is partitioned into disjointed groups. Thus, the analysis of world trade flows in 1995 led geographer Jessie Poon to conclude that the USA–Europe–Japan triad taught in school textbooks was a myth and that the world was divided into four regions, at least two of which shared the "European" space (Poon et al. 2000).

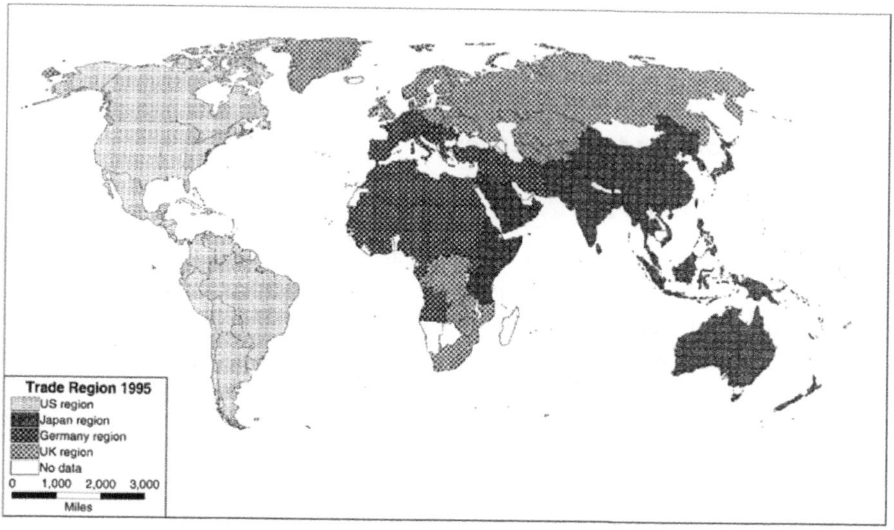

Figure 10.4. *Division of the world into regions based on trade flows in 1995 (source: Poon et al. (2000))*

10.3.1.4. *Europe in people's minds: the region as a social representation*

Finally, we also find works that, privileging the social representations of space, highlight the diversity of discourses and practices of Europe as a perceived, lived space, possibly as an appropriate territory. Some analyze practices from below and popular representations. Others (high geopolitics) are interested in the practices and representations of Europe by decision-makers – politicians, advisors, civil servants, lobbyists, activists – or other normative bodies such as the media. Unlike the field of popular geopolitics, which favors a documentary analysis based on discourse analysis (Hagen 2003; Müller 2011) closer to geohistorical analysis, this work proceeds through the analysis of surveys conducted among targeted populations such as students, ESPON project actors (Didelon 2010; Didelon et al. 2011, 2018; Toureille 2017), more rarely stratified surveys (Didelon and Grasland 2021), using mental maps sometimes associated with elements of semantic description (e.g.

words associated with the term "Europe" (Brennetot et al. 2013)). One of the objectives is often the comparison of representations between study countries.

These different currents, which consider the notion of "Europe" from a heuristic perspective, are most often complementary. From a formal point of view, they all share an affinity for cartographic visualization, whether as an object of analysis (documentary analysis) or a method of investigation (exploratory cartography, mental maps). If they take a regional dimension, they do not all explicitly call upon the concept of region and even less upon the field of regional studies.

10.3.2. *Limitations and perspectives*

10.3.2.1. *A focus on the EU*

Many works remain focused on the EU as an object of study and on the reading grids induced by the planning objectives of the EU and its territorial agenda. If we take the example of work on the social representations of the EU, contributions by geographers are rare: we must often look to political scientists (Chaban 2008, 2013; Cautrès and Monceau 2011). When they do exist, they often focus on surveys conducted in a limited number of the richest or most powerful member states (Anholt 2010, 2011).

10.3.2.2. *The influence of economics and political science*

Many of these works remain dependent on a field of European studies where geographers occupy a minority position compared to economists and political scientists, the two disciplinary affiliations claimed by the international reference journal: *Journal of European Market Studies*. Lawyers are also active in this field and have recently been making more contacts with geographers (Brunet et al. 2019). This extra-disciplinary tropism can therefore lead to some frustrations (lack of critical analysis of the geographical framework used to define Europe, use of space through the convocation of grids mainly used for the analysis of indicators, often statocentric perspectives).

10.3.2.3. *The weak influence of the critical turn in human geography*

Unlike in the anglophone context, French regional studies devoted to Europe have not borrowed the critical current impelled in particular in the sphere of critical political geography (for an overview of European studies in anglophone geography, see Bachmann (2021)). A few notable exceptions exist, notably on the theme of migration geography and the EU's external borders (Bernardie-Tahir and Schmoll 2015; MIGREUROP 2017; Schmoll 2018).

Consequently, the majority of the French works cited above privilege the representations of elites, particularly the political actors of states and intergovernmental organizations. This remark is valid for works in political geography and geohistory, which often rely on documents drawn from the culture of elites (works of art, scholarly cartography, for example). In this context, it is probably more difficult to see alternative or unexpected representations of "Europe" emerge.

10.3.2.4. *Few systemic analyses crossing social and spatial diversity of points of view*

Representational approaches are often limited to the exploration of a single dimension: the international comparison of the same type of discourse or the confrontation of different discourses in the same territory. In geography, few works have attempted a systematic analysis of these two dimensions. In political science, for example, Lucarelli (2014) highlighted the discrepancy between the representations of actors (citizens, government elites, civil society groups and media) located in member countries, and others in non-member countries where the EU is identified as a power and a potential economic partner and also criticized for its "condescending attitude", its "European protectionism" leading to its perception as a "neo-colonial power". The idea is no longer only to work on the image of Europe seen from the outside or the inside, but to study the reciprocal representations of Europeans and their neighbors by considering actors located at different levels. It is also possible to consider the analysis of regionalization proposed by actors who are more distant from the spheres of decision-making or marginalized, such as LGBTQ communities in Eastern Europe, or immigrants, for example.

While researchers involved in each of these approaches meet occasionally in the context of publications and projects, there is little space for exchange and joint work. The fragmentation of European studies in geography, as well as the highly variable institutionalization of the discipline from one country to another, undoubtedly explain the often interdisciplinary openness of the work envisaged on these questions.

10.3.3. *Some ideas for reflection: returning to the deeper meaning of regionalization*

Starting from this assessment – specifically francophone – how can we specify the conceptual foundations of a regional geography more adapted to the study of Europe, this complex geographical object?

10.3.3.1. *Regionalize: produce statements*

It is important to return to the deeper meaning of regionalization, an intellectual exercise with heuristic value that is necessarily subjective:

> [T]he region is not a 'natural' or 'absolute' entity, which would exist independently in the real world and would only wait to be discovered, it is rather an intellectual construction – a 'construct' – that exists only in the mind of the 'regionalizer' […] [R]egionalization is the act of dividing a space into logical geographical areas or regions. The specific system of regions that results from this process depends on the set of criteria one applies (Coffey 1995, p. 150).

There are two possible strategies for producing regions: cutting a predefined set into subsets; or trying to define the boundary of a defined set within a larger set. The second part of the quotation is particularly important in order to avoid falling into the trap of a regionalization that would consist of essentializing a space of political construction and thus contribute to the legitimization of ideologies. Given its eminently ideological character and its high political charge, regionalizing "Europe" therefore requires a critical approach to the ensembles analyzed or constructed as a result of the regionalization process.

10.3.3.2. *What geometry to give to the region?*

A region is a portion of the earth's space defined by its singularity, and most geographers agree that it is a contiguous portion of space, even though it may include some exclaves or enclaves (Richard and Marei 2020). However, in the case of "Europe", this idea is not obvious. Given its past, especially its colonial past, Europeans have shaped the world in their own image and built a number of "projected Europes" (Lévy 1997), corresponding in particular to the colonies. This proximity can be seen, for example, in architecture: Buenos Aires, Algiers and Washington were the playgrounds of French urban planners. Beyond colonization alone, the paradigm of modernity has contributed to the Europeanization of the world, making the characteristics of this "situated modernity" synonymous with progress and the university, as highlighted by the subalternist critique calling for the "provincialization of Europe" (Chakrabarty 2000; Amselle 2004). More concretely, a number of ultra-marine territories controlled by European states are sometimes very far from their historical metropolises. As such, the French overseas departments and regions, as well as the Portuguese islands of the Azores and Madeira or the Spanish Canary Islands benefit from the status of outermost regions of the EU according to Article 349 of the TEU. Should the confetti of empires, sometimes tens of hours' flight from the capital, be excluded from "Europe"? Very few studies have so far questioned the "Europeanness" of these territories: if Europe is a territory characterized by a common experience, how is it perceived, experienced and appropriated by the populations of these territories?

10.3.3.3. *Objectivize the definition criteria?*

The main problem with the various delimitations of Europe seems to reside, in the end, in the absence of explicit criteria and procedures for including or excluding this or that portion of space (Montello 2003). Are we seeking to confirm a delimitation conceived a priori by retaining only those criteria that comfort common sense, or are we prepared to take the risk of creating a European region with unexpected contours?

– A *morphological criterion* based solely on the distribution of land and sea could support visions of a Eurasian or Eurasian space, but would certainly not coincide with the territory of the EU if applied objectively.

– A *homogeneity criterion* combining – for example – a high level of wealth per capita, a significant aging of the population, the abolition of the death penalty, and two or three other ad hoc criteria would lead without difficulty to the more or less exact delimitation of the current EU territory, with a few exceptions.

– A *functional criterion* using, for example, trade flows or foreign direct investment could lead to a "Greater Europe" enlarged to include its southern and eastern neighbors, even though this space is likely to evolve over time and cannot be based on fixed limits.

– A *strategic criterion* based on UN votes or military alliance treaties would result in a new, possibly transatlantic region involving the EU and North America.

But more importantly, the *subjective criteria of the representations* that mix each of the above ingredients would lead to yet another result, variable from one EU country to another and even from one individual to another, and not fixed in time!

10.4. Conclusion: the geography of Europe, a psychoanalytical exercise for Europeans

If the geography of Europe is a puzzle, perhaps it is above all when it is the work of Europeans. One of the difficulties of the exercise is to have the necessary critical distance for the study of an object which, for researchers located in Europe, is a lively and rarely consensual political subject.

10.4.1. *A necessarily plural and complex object*

As seen by Europeans, Europe is rarely presented as a simple object. When works on Europe are published, it is frequently with the addition of qualifiers breaking it down into multiple Europes covering various supranational spaces: Scandinavia, the Baltic region, the Balkans, Central Europe, Eastern Europe, the Middle East, the Mediterranean, the East, the North (Escach 2014; Escach and Vaudor 2014; Von Hirschhausen 2017; Guillaume 2020). This division of Europe into subsets is also found in the universal geographies published in France over the last two centuries, where Europe is never considered as a whole, but as an assembly of regional subsets. This vitality of the divisions can be explained by their existence as objects of study for a long time structured by multidisciplinary research teams developing an approach by "cultural areas". This complexity is perhaps especially specific to a European point of view. For a European, it will perhaps be easier to consider as a set of distant territories, even though they are diverse (Oceania, for example), as a clearly identified region. On the contrary, their greater knowledge of a Europe that they live in and that they have perhaps traveled through will lead them to the observation of diversity.

The divisions of Europe also highlight its national visions. A French geography of Europe gives pride of place to France and can bring it closer to the Overseas Territories, North Africa and the Mediterranean for reasons linked to its colonial history (Deprest 2002). This would not be the case for Germany, the United Kingdom or Turkey.

10.4.2. *Dividing the world into regions: a way to find our place in it*

The geographies of Europe raise the question of the otherness it faces, and that of the place of Europe and Europeans in the world. Demangeon (1920) proposed comparisons with other world centers (Japan, the United States). The recurrent concern about Europe's supposed loss of influence in relation to other emerging centralities can be read in this work. A few years later, the 1929 crisis played a central role in the awareness of interdependencies between national economies and of the globalization underway (Arrault 2007). How can we not see this type of concern in EU commissions for reports on the place of "Europe in the world", or when scientists, the media and the general public underestimate the inescapable role of the new centers of the world economy such as China, India and the Arab-Persian Gulf countries (Claval 2014). Finding Europe in the world and reflecting on its limits would thus lead Europeans to talk about themselves and respond to concerns about their possible loss of influence. These concerns are not limited to the economic sphere alone, they raise the question of the historical role played by

Europeans in the world through colonization and the diffusion of socio-political models (the question of modernity). In his time, Demangeon asked: "How did the European present himself to the other peoples of the world?" Siegfried answered bluntly: "That the white man has behaved like a slave trader, a pirate, a greedy and insatiable exploiter, is, alas, only too true" (Siegfried 1935). Like a psychoanalysis for the authors who attempt the exercise, carrying out a geography of Europe leads to questioning the inferiority and superiority complexes of Europeans, and thus to reflecting on their place in the rest of the world, the study of the world and of the relations between Europe and the other regions becoming inseparable from the analysis.

10.5. References

Ahnström, L. (1993). Europe: Culture area, geo-ideological construct or illusion? *Norsk Geograpfish Tidsskirft/Norwegian Journal of Geography*, 47(2), 57–67.

Amin, S. (1988). *L'Eurocentrisme, critique d'une idéologie*. Anthropos, Paris.

Amselle, J.-L. (2004). Le retournement du Monde. Ou les enjeux d'une déconstruction de l'Occident. *Lignes*, 15, 155–170.

Anholt, S. (2010). Definitions of place branding – Working towards a resolution. *Place Branding and Public Diplomacy*, 6(1), 1–10.

Anholt, S. (2011). Beyond the nation brand: The role of image and identity in international relations. *Exchange. The Journal of Public Diplomacy*, 2(1), 6–12.

Arrault, J.-B. (2007). Geography and globalization. French geographers facing the 1930's crisis. *Annales de géographie*, 657(5), 493–512.

Bachmann, V. (2021). Visionary geographies and European studies. *Progress in Human Geography*, 46(2), 416–440.

Bafoil, F. (2006). Resumen. *Revue internationale de politique comparée*, 13(2), 213–238.

Bafoil, F. (2008). L'européanisation : héritages, résistances, mobilisations collectives et cohésion. *Critique internationale*, 41(4), 159–169.

Bailleul, H., Baudelle, G., Josselin, J.-P. (2019). Editorial : les petites villes européennes comme enjeu d'équité territoriale. *Belgeo. Revue belge de géographie*, 3 [Online]. Available at: http://journals.openedition.org/belgeo/35110.

Baron, M., Sabot, E., Grasland, C. (2010). *Villes et régions européennes en décroissance : maintenir la cohésion territoriale ?* Hermès Lavoisier, Paris.

Beauguitte, L. (2011). L'assemblée générale de l'ONU de 1985 à nos jours : acteur et reflet du système-monde. Essai de géographie politique quantitative. PhD Thesis, Université Paris-Diderot, Paris VII.

Beckouche, P. (2008). *Les régions nord-sud : Euromed face à l'intégration des Amériques et de l'Asie orientale.* Belin, Paris.

Beckouche, P. and Richard, Y. (2008). *Atlas d'une nouvelle Europe : l'UE et ses voisins, vers une région mondiale ?* Autrement, Paris.

Beckouche, P. and Richard, Y. (2013). *Atlas de la grande Europe : économie, culture, politique.* Autrement, Paris.

Beckouche, P., Grasland, C., Didelon, C., Richard, Y., Zanin, C. (2008). *Europe in the World. Territorial Evidence and Visions.* Université Paris 1, Université Paris 7, CNRS.

Beckouche, P., Besnard, P., Pécout, H. (2016). *Atlas of Challenges and Opportunities in European Neighbourhoods.* Springer, New York.

Bernardie-Tahir, N. and Schmoll, C. (2015). Iles, frontières et migrations méditerranéennes : Lampedusa et les autres. *L'Espace Politique*, 25, 1.

Beyer, A. (2020). Les institutions internationales, révélatrices des dynamiques régionales. Pour une lecture géographique de l'espace aérien européen et de son intégration. *Belgeo. Revue belge de géographie*, 4 [Online]. Available at: http://journals.openedition.org/belgeo/43572.

Boulineau, E. and Bonerandi, E. (2014). *La pauvreté en Europe : une approche géographique.* Presses Universitaires de Rennes.

Bourdin, S. (2020). Concevoir autrement la politique de cohésion de l'Union européenne : la spécialisation intelligente comme approche davantage territorialisée ? *Économie et institutions*, 18, 28 [Online]. Available at: http://journals.openedition.org/ei/6608.

Brennetot, A. and Rosemberg, M. (2013). Géographie de l'Europe et géographie de la construction européenne. Les images contradictoires de l'Europe et de l'Union européenne dans quelques manuels de géographie contemporains. *L'Espace Politique*, 19, 1.

Brennetot, A., Emsellem, K., Guérin-Pace, F., Garnier, B. (2013). Dire l'Europe à travers le monde. Les mots des étudiants dans l'enquête EuroBroadMap. *Cybergeo: European Journal of Geography*, 630. doi: 10.4000/cybergeo.25684.

Briot, N. (2021). Villes en réseaux. Les coopérations internationales des villes françaises : spatialisation, internationalisation, européanisation. PhD Thesis, ENS de Lyon, Lyon.

Brunet, S., Lebon, L., Richard, Y. (2019). Prolifération des territoires et représentations de l'Union européenne. *Annales de Droit*, Presses Universitaires de Rouen et du Havre.

Capdepuy, V. (2018). *50 histoires de mondialisations : de Neandertal à Wikipédia.* Alma éditeur, Paris.

Cautrès, B. and Monceau, N. (2011). *La Turquie en Europe. L'opinion des Européens et des Turcs.* Presses de Sciences Po, Paris.

Chaban, N. and Holland, M. (2008). *The European Union and the Asia-Pacific: Media, Public and Elite Perception of the EU.* Routledge, London.

Chaban, N. and Holland, M. (2013). Special issue: Changing external perception of the EU? Vision from the Asia-Pacific. *Baltic Journal of European Studies*, 3(3). doi: 10.2478/bjes-2013-0018.

Chakraparty, D. (2000). *Provincializing Europe. Postcolonial Throught and Historical Difference*. Princeton University Press.

Claval, P. (2014). Le déclin de l'Europe. Un problème et sa perception par les géographies. *BSGLg, Bulletin de la Société Géographique de Liège*, 62(1).

Coffey, W. (1995). Géographie, économie, science régionale. In *Encyclopédie de Géographie*, Bailly, A., Ferras, R., Pumain, D. (eds). Economica, Paris.

Demangeon, A. (1920). *Le déclin de l'Europe.* Payot, Paris.

Demangeon, A. (1932). Les conditions géographiques d'une union européenne. Fédération européenne ou ententes régionales ? *Annales*, 4(17), 433–451.

Deprest, F. (2002). L'invention géographique de la Méditerranée : éléments de réflexion. *L'Espace géographique*, 31(1), 73–92.

Didelon, C. (2010). A vision of Europe. The drawing of Europe by researchers implied in ESPON. *Annales de géographie*, 673(3), 211–228.

Didelon, C. and Grasland, C. (2014). Internal and external perception of Europe/the EU in the world through mental maps. In *Communicating Europe in Times of Crisis. External Perceptions of the European Union*, Chaban, N. and Holland, M. (eds), Palgrave-Macmillan, Basingstoke-New York.

Didelon, C. and Grasland, C. (2020). L'union européenne à la carte. In *Un panel français. L'étude longitudinale par internet pour les sciences sociales*, Duwez, E. and Mercklé, P. (eds). INED Editions, Paris.

Didelon, C., Grasland, C., Richard, Y., Sillère, G. (2008). *Atlas de l'Europe dans le monde.* Reclus CNRS GDRE, La Documentation française, Montpellier, Paris.

Didelon, C., Ruffray, S., de, Boquet, M., Lambert, N. (2011a). A world of interstices: A fuzzy logic approach to the analysis of interpretative maps. *The Cartographic Journal*, 48(2), 100–107.

Didelon, C., Richard, Y., Van Hamme, G. (2011b). *Le territoire européen.* Presses Universitaires de France, Paris.

Didelon-Loiseau, C., de Ruffray, S. Lambert, N. (2018). Mental maps of global regions: Identifying and characterizing "hard" and "soft" regions. *Journal of Cultural Geography*, 35(2), 210–229.

Dollfus, O. (2001). *La mondialisation.* Presses de Sciences Po, Paris.

Escach, N. (2014). Réseaux de villes et recompositions interterritoriales dans l'espace baltique. PhD Thesis, Lyon 2.

Escach, N. (2015). Les réseaux de villes baltiques : la dimension spatiale de la recomposition des niveaux. *L'information géographique*, 79(3), 34–53.

Escach, N. and Vaudor, L. (2014). Réseaux de villes et processus de recomposition des niveaux : le cas des villes baltiques. *Cybergeo: European Journal of Géography*. doi: 10.4000/cybergeo.26336.

Faludi, A. (2004). Territorial cohesion: Old (French) wine in new bottles? *Urban Studies*, 41(7), 1349–1365.

Faludi, A. (2007). The European model of society. In *Territorial Cohesion and the European Model of Society*, Faludi, A. (ed.). Lincoln Institute of Land Policy, Cambridge, MA.

Gana, A. and Richard, Y. (eds) (2014). *La régionalisation du monde : construction territoriale et articulation global-local*. IRMC, Tunis.

Gourdon, P. (2021). La coopération entre villes européennes : convergences dans l'action publique urbaine par la circulation transnationale de modèles. PhD Thesis, Université Paris 1 Panthéon-Sorbonne.

Grasland, C. and Hamez, G. (2005). Vers la construction d'un indicateur de cohésion territoriale européen ? *L'Espace géographique*, 34(2), 97–116.

Grasland, C. and Van Hamme, G. (2010). La relocalisation des activités industrielles : une approche centre-périphérie des dynamiques mondiale et européenne. *L'Espace géographique*, 39(1), 1–19.

Grataloup, C. (2007). *Géohistoire de la mondialisation : le temps long du monde*. Armand Colin, Paris.

Grataloup, C. (2009). *L'invention des continents : comment l'Europe a découpé le monde*. Larousse, Paris.

Guillaume, J. (2020). Entre régionalisme et régionalisation, le difficile positionnement du Norden en Europe. *Belgeo. Revue belge de géographie*, 4 [Online]. Available at: http://journals.openedition.org/belgeo/43803.

Hagen, J. (2003). Redrawing the imagined map of Europe: The rise and fall of the "center". *Political Geography*, 22(5), 489–517.

von Hirschhausen, B. (2017). Leçon des frontières fantômes : les traces du passé nous viennent (aussi) du futur. *L'Espace géographique*, 46(2), 97–105.

Jean, Y. and Baudelle, G. (2009). *L'Europe – Aménager les territoires*. Armand Colin, Paris.

Jessop, B. (2002). Liberalism, neoliberalism, and urban governance: A state–theoretical perspective. *Antipode*, 34(3), 452–472.

Laménie, B. (2019). La dimension urbaine de la politique européenne de cohésion et la construction métropolitaine : le cas de l'Ile-de-France. PhD Thesis, Université de Paris.

Le Texier, M. (2014). Les circulations internationales en Europe au XIXe siècle : de l'étude des circulations monétaires à la modélisation d'un système complexe. PhD Thesis, Université Paris Diderot, Paris 7.

Lévy, J. (1997). *L'Europe : une géographie*. Hachette, Paris.

Lewis, M.W. and Wigen, K. (1997). *The Myth of Continents: A Critique of Metageography*. University of California Press, Berkeley, CA.

Lucarelli, S. (2014). The EU's leadership in global governance: Perception from the others. In *Communicating Europe in Times of Crisis. External Perceptions of the European Union*, Chaban, N. and Holland, M. (eds). Palgrave-Macmillan, Basingstoke-New York.

Mareï, N. and Richard, Y. (2018). *Dictionnaire de la régionalisation du monde*. Atlande, Paris.

Mareï, N. and Richard, Y. (2020). "Regards". Régionalisations du monde et intégration (macro)régionale. Pour une montée en généralité géographique. *Belgeo. Revue belge de géographie*, 4 [Online]. Available at: http://journals.openedition.org/belgeo/43451.

Migreurop (2017). *Atlas des migrants en Europe : approches critiques des politiques migratoires*, 3rd edition. Armand Colin, Paris.

Montabone, B. (2013). *L'Union européenne et la Turquie : les enjeux d'un développement régional*. Presses Universitaires de Rennes.

Montello, D. (2003). Region in geography: Process and content. In *Foundation of Geographic Information Science*, Duckham, M., Goodchild, M.F., Worboys, M. (eds). Taylor & Francis, London.

Müller, M. (2011). Doing discourse analysis in critical geopolitics. *L'Espace Politique. Revue en ligne de géographie politique et de géopolitique*, 11 February, 2010-3.

Palle, A. (2017). L'Espace énergétique européen : quelle(s) integration(s) régionale(s) ? : réseaux, normes, marchés, politiques, des intégrations à plusieurs échelles ? PhD Thesis, Université Paris 1 Panthéon-Sorbonne.

Parker, W.H. (1960). Europe: How far? *The Geographical Journal*, 126(3), 278–297.

Poon, J.P., Thompson, E.R., Kelly, P.F. (2000). Myth of the triad? The geography of trade and investment "blocs". *Transactions of the Institute of British Geographers*, 25(4), 427–444.

Radaelli, C.M. (2004). Europeanisation: Solution or problem? Report, Social Science Research Network, Rochester, NY. SSRN Scholarly Paper, No. ID 601163.

Richard, Y. (2008). L'Union européenne et ses voisinages. Organiser la relation pour construire une région mondiale. In *Union européenne – Voisinages, la quête d'une intégration réigionale*, Richard, Y. (ed.). Strates, Ladyss, Paris.

Schmoll, C. (2018). Quoi de neuf sous le soleil de Sicile ? Migrations internationales et enjeux géopolitiques en mer Méditerranée (2011–2017). *EchoGéo*, 14, 02.

Siegfried, A. (1935). *La crise de l'Europe*. Calmann-Lévy, Paris.

Toureille, E. (2017). La Turquie en Europe, la Turquie dans le monde : analyse d'une catégorie macro-régionale au regard d'une géographie des représentations de l'Europe et du monde turco-centrée. PhD Thesis, Sorbonne Paris Cité.

Toureille, E. (2020). Saisir l'"Europe" à la lueur du discours officiel de l'UE : l'analyse de données au service d'une régionalisation critique de l'Europe. *Belgeo. Revue belge de géographie*, 4 [Online]. Available at: http://journals.openedition.org/belgeo/43877.

Valéry, P. (1919). *La crise de l'esprit*. Nouveau Roman Français. Paris.

Vandermotten, C. and Dézert, B. (2008). *L'identité de l'Europe : histoire et géographie d'une quête d'unité*. Armand Colin, Paris.

Wallerstein, I.M. (2009). *Comprendre le monde : introduction à l'analyse des systèmes-monde*. La Découverte, Paris.

11

Three Exercises in the Regionalization of Europe

Antoine LAPORTE[1], Etienne TOUREILLE[2] and Claude GRASLAND[3]

[1] *UMR 5600 EVS, École Normale Supérieure de Lyon, France*

[2] *UMR 6266 IDEES, Université de Rouen – Normandie, France*

[3] *UMR 8504 Géographie-cités, Université de Paris, France*

11.1. Introduction: the search for Europe and three tracks for its quest

The definition (or indefinition) of "Europe" is at the heart of permanent debates within the scientific sphere as well as in the political world. The object is neither determined in its perimeter, nor in what would make its identity, specificities or institutional attributions. The confusion between the common term Europe and the European Union (EU) increases the difficulties in defining this part of the world and in studying the discourses and representations of which it is the object (Bachmann 2022).

The identification of Europe with a larger whole than the state is due to the advanced interweaving of national territories in terms of the circulation of wealth, linguistic practices and international mobility. More than ever, it is in Europe, despite the fact that it was the cradle of the Westphalian state system, that the state level seems to have delegated the most of its sovereignty to other levels, while being less and less an indestructible territorial framework in both individual practices and collective representations. This is partly due to borders that are often being erased and to the coexistence of states whose proliferation, revealing a fragmented pavement, is unique in the world. There are almost 50 of them between Azerbaijan and Iceland, whose capitals are closer than the arc joining Vancouver Island and

Regionalization of the World,
coordinated by Pierre BECKOUCHE and Yann RICHARD.

Newfoundland Island, both of which belong to Canada. And unlike other large regions of the world, no state is in a position of supremacy over this group as are the United States in North America or, to a lesser extent, Brazil in South America.

The institutional inclusion of European states in supranational structures (the EU, the Nordic Council, the Commonwealth of Independent States) is complex. It is compounded by alliances of varying geometry in areas considered to be within the traditional domain of sovereignty, whether it be border crossing rules (Schengen Agreements), currency (the Eurozone) or defense. To take only the example of border crossings, agreements on movement between states produce new political territories. Within the EU, they have generated extended spaces where international limits no longer form a real barrier, while outside they can lead to outright closures, in particular to international migration, producing numerous dramas. Brexit is a textbook case of the international political tensions that can be generated by a recomposition of these spaces of circulation, particularly in Ireland.

The exercise of dividing the world into large regions does not take place according to simple partitions, but is rather a matter of fuzzy geometry because of the frequent cases of multiple memberships. One can be both a "European" and a "Mediterranean" state, part of the "Near East" or of the "former Soviet bloc", as many intersecting entities with often unclear perimeters. This involvement of European states is all the less coherent a priori since, through their history, many of them have exercised forms of hegemony over large parts of the continent or the world (Badie 2019), notably via the experience of colonial empires with original forms of networks of territories scattered over the world maps.

We can see the fragility of the concept of "European power", which would suppose the existence of a delimited space with an identity and an autonomous capacity for action. The addition of the power of neighboring states, but with little coordination of their policies, leads to a result where the whole is *less than* the sum of its parts.

Indeed, community integration has left a deep mark on European territories. However, it does not prevent the significant development of nationalist populism. In many European countries, we are seeing the electoral success of parties that advocate withdrawal rather than greater European integration. The vote in favor of Brexit, the greater presence of the far right in the European Parliament since 2014 as well as in the national chambers, leads us to believe that in the minds of many, the description of a European geographical entity should not be accompanied by a gradual transformation into a transnational European territory or federation. In the vicinity of the community territory, Russia, Turkey or, at its level, Egypt exercise both nationalistic and imperialist policies, advancing their pawns with a statocentric grammar of their regional power claims.

Beyond the question of power and its projection, the existence of a Europe requires a set of multiple memberships of the states, which contributes to their integration, at least symbolically, into a regional whole. This is both institutional and economic. It refers to the place given to the expression "Europe" in discourse and to the regional assignment of states. In these areas, the term "European" or "community" (which implicitly refers to the EU) is in competition with strong regional groupings that are just as politically significant.

This chapter proposes three empirical illustrations of Europe through the systems of relations between states in a space that we will call European, which implicitly questions the coherence of the whole in which they are inserted. The first illustration deals with the position of European states within diplomatic relations through the establishment of embassies. It aims to measure the power projection capacity of European states in relation to other chancelleries in the world. The second illustration proposes a reflection on Turkey's membership in supranational entities, a state in an interface position between an integrated Europe and other regional spaces invested by Turkish diplomacy, such as the Black Sea area or the Turkish-speaking world of Central Asia. The third illustration poses the links between European states and macro-regional geographical objects in the text of highly consulted newspapers, considered as references, in the French media landscape.

11.2. Embassy networks in the world and the renewed place of European states

The diplomatic presence directly poses the existence of a European region because it allows us to see, beyond a measure of the influence of European states, forms of regional integration. International relations, which are materialized by the existence of places such as embassies, are considered as a pillar of national sovereignty and power.

Diplomatic relations are the set of links established between the governments of states. In a world thought as a mosaic of states, diplomacy is the agent that smooths the relations between entities considered as sovereign, the states.

Interest in diplomatic apparatuses has developed in the literature since the 1960s (Brams 1966; Xierali and Liu 2006; Neumayer 2008; Beauguitte and Didelon 2011). The number and distribution of embassies in the world are indicators of power (real or desired) because they reveal the investment, sometimes considerable, made by states in this area. Databases on the subject are now available to provide an overview, such as those of the Correlates of War Project (COW) founded at the University of Michigan or those of the Lowy Institute, a think tank based in Sydney. Being more difficult to handle and little used, articles in the standardized form of the online encyclopedia *Wikipedia* allow for an exhaustive and up-to-date view.

According to the data collected by COW, in 2005, if we took two of the 192 sovereign states listed by the UN at random, we had only a 22% chance of finding countries whose relations were based on an exchange of embassies. During the 20th century, this percentage decreased as the number of states increased and therefore the burden for each state to have a significant diplomatic apparatus increased. The absence of a resident embassy does not prevent regular diplomatic relations. It does, however, hinder the capacity for discussion, influence and representation. If we add to this the fact that the missions of embassies are not limited to political relations alone, but also include the economy, culture, research or defense, these structures can be several dozen people strong and have considerable budgets.

While the absence of diplomatic representations of a given state in a given capital sometimes reveals geopolitical tensions (there is no US embassy in Iran, for example), in the vast majority of cases, deployment capacities depend directly on public wealth (Table 11.1). States with modest finances will often have no more than 30–50 embassies in the world and will form "weak diplomatic offices" (Beauguitte and Didelon 2011). These states prefer to have a presence in a few major powers and with their neighbors, with whom border issues, migration questions and privileged trade links are essential elements of discussion.

The ranking of the most important diplomatic apparatuses reveals a hierarchy of power. At the beginning of the 20th century, the main industrial and colonial powers, primarily France and Great Britain, had the most extensive embassy networks. In 1914 and 1940, even European states of more modest economic size were still among the most represented in the world, such as Spain, Belgium, Portugal and Denmark. Already powerful before the war, the United States then asserted itself in the context of the Cold War as the world's leading diplomacy, while the USSR and China rose to 5th and 6th places, respectively. But since 2017, the United States has been supplanted by China. The position of European states is declining. Between 1985 and 2019, France fell from 2nd to 3rd, Germany from 3rd to 6th, and Italy from 7th to 11th. At the same time, emerging states have established extensive diplomatic networks. Turkey opened 59 representations between 2005 and 2019, Brazil 52 representations, South Korea 31 representations and China 26 representations. Europe therefore no longer has a monopoly on large diplomatic apparatuses.

	1914		1940		1985		2005		2019	
	Country	Nb	Country	Nb	Country	Nb	Country	Nb	Country	Nb
1	France	38	France	55	United States	136	United States	162	China	169
2	Germany	37	United States	52	France	123	France	145	United States	168
3	Great Britain	37	United Kingdom	51	West Germany	118	China	143	France	161
4	United States	37	Belgium	49	United Kingdom	115	United Kingdom	141	United Kingdom	152
5	Italy	35	Italy	45	USSR	107	Germany	140	Japan	151
6	Spain	34	Germany	44	China	100	Russia	130	Germany	150
7	Austria-Hungary	31	Netherlands	41	Italy	97	Japan	120	Russia	144
8	Belgium	30	Argentina	40	Japan	96	Italy	116	Turkey	140
9	Russia	28	Sweden	38	India	88	Egypt	112	Brazil	138
10	Netherlands	26	Denmark	36	Belgium	87	India	104	Egypt	126
11	Brazil	25	Brazil	35	Egypt	85	Canada	95	Italy	124
12	Argentina	24	Chile	34	Brazil	77	Spain	92	India	123
13	Mexico	21	Mexico	34	Canada	76	Brazil	86	Spain	115
14	Portugal	21	Norway	33	Spain	75	Belgium	85	South Korea	114
15	Chile	20	Spain	33	Yugoslavia	74	Netherlands	84	Netherlands	107

Table 11.1. *The main diplomatic representations in the World. Source: COW (Correlates of War) database, Global Diplomacy Index and authors' calculations*

However, European states are rarely weak diplomats. Only 5 of the 27 EU members and Iceland have less than 50 embassies in the world. The downgrading of the smallest European states must also be put into perspective. For if the installation of embassies in the world constitutes the most visible part of diplomatic power, the capacity to host embassies at home is just as important. In this area, European states benefit from the fact that there is nowhere else in the world where there are so many states in such a small area. In a context of open international movements in the Schengen area, many chancelleries consider that an embassy can be located in one capital without worrying and be accredited for several others, in neighboring and easily accessible states. Almost every state in the world is represented at least three times in the European Union. Brussels, the seat of the main EU institutions, has become the city of diplomatic residence for 180 states of the world, a level roughly equal to that of Washington and Beijing.

Two elements finally argue for an approach of the EU (even more than "Europe") as a global center of diplomatic activity. Firstly, EU legislation indicates that any EU citizen can find help wherever they are in any of the member states' embassies if their state is not directly represented. Secondly, since the adoption of the Lisbon Treaty in 2007, the EU has had a European External Action Service, a diplomatic structure parallel to that of the states. Its missions do not have the status of embassies, but its location is now comparable to that of the most powerful diplomats.

11.3. Europe through the titles of regional IGOs: a perspective from Turkey

The geometry of the relations of inclusion, exclusion or intersection between the geographical objects claiming to be part of Europe seems complex, but it is finally more organized than it appears at first sight. As an exercise, we propose the example of Turkey by confronting two types of Intergovernmental Organization (IGO): those that include the word "Europe" in their name (whether Turkey is a member or not) and those of which Turkey is a member (whether the word "Europe" is present or not). We can then examine the question of Turkey's "Europeanness" by looking at the extent to which the two sets of organizations overlap or not.

Leaving aside organizations with a cultural or sporting vocation (Eurovision, European soccer or basketball leagues, for example), Turkey maintains strong and lasting relations with at least three IGOs with a European vocation: the EU, the Council of Europe (CoE) and the Organization for Security and Cooperation in Europe (OSCE). It is therefore possible to analyze Turkey's place in relation to these different organizations, which, when superimposed, make it possible to identify a conceptual core of countries belonging to a political "Europe" defined through these titles and the fringes of membership (Figure 11.1). The result is an aureole organization that can be analyzed as a vague whole.

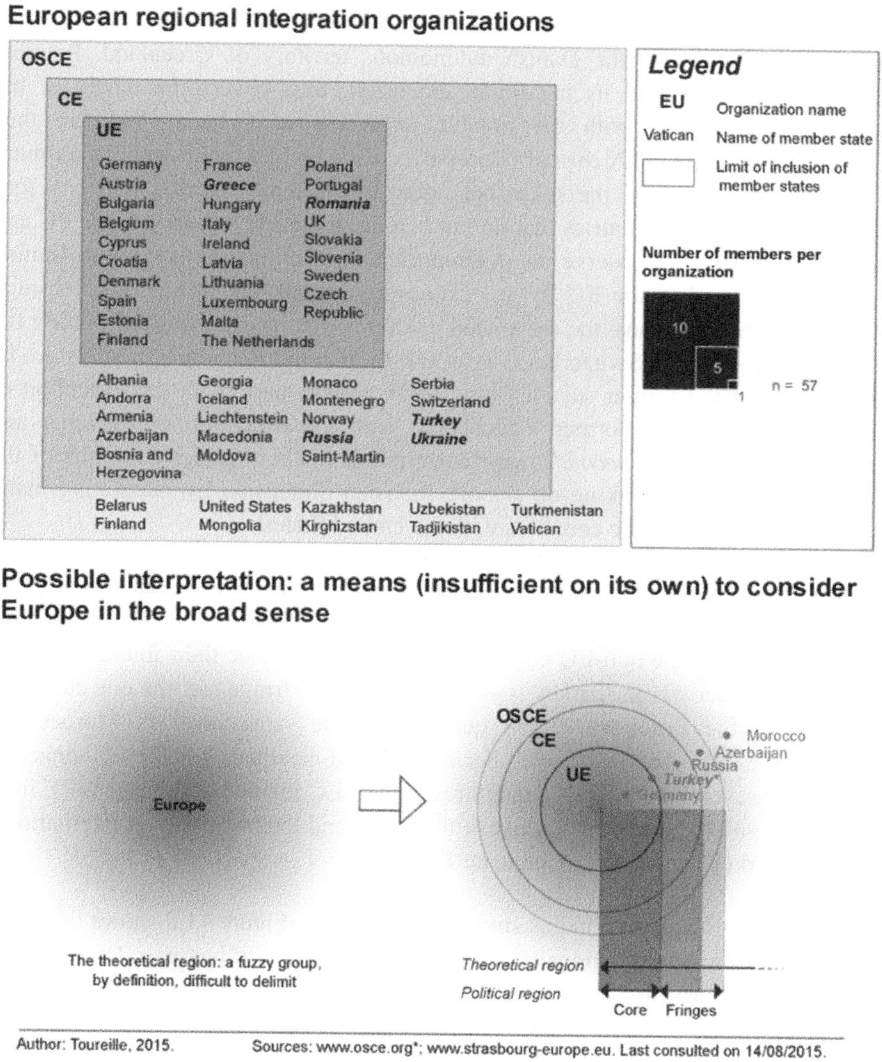

Figure 11.1. *Political "Europe" captured through the titles of IGOs with a European focus. Source: Toureille 2017. For a color version of this figure, see www.iste.co.uk/beckouche/regionalization.zip*

The extent of "Europe" captured through the three IGOs is relatively broad. In its maximum extension, it takes on a distinctly Eurasian dimension: it includes Russia and a whole set of territories of the former Soviet Union, right up to the borders of Central Asia with countries bordering Iran, Afghanistan and even China. In the west, it includes Iceland and the Danish autonomous territory of Greenland. Political "Europe", understood in its maximum extension, thus covers the whole of the Northern Hemisphere, with the notable exception of Canada! We can thus distinguish three groups. A *core* of "Europe" is made up of a group with a maximum probability of belonging: the states belonging to the three selected IGOs. A *first fringe* corresponds to countries that do not belong to the EU but only to the EC and the OSCE, where we observe an overrepresentation of micro-states with limited sovereignty, linked to the EU by association agreements (Andorra, Liechtenstein, Monaco and San Marino), the associated countries that have renounced joining the EU (Iceland, Norway, Switzerland, to which the United Kingdom should now be added), part of the former socialist bloc in its eastern part (with the exception of Belarus), the states of former Yugoslavia, with the exception of Slovenia and Croatia, and Turkey. A *second fringe* corresponds to the member countries of the OSCE alone, which is taking on a more Eurasian dimension by integrating many Central Asian and Atlantic countries with the United States.

This first fuzzy regionalization of a political "Europe" shows that focusing the reasoning on the EU – the most restrictive geometry – is not neutral. The EC and the OSCE, to which some non-EU countries have belonged since their foundation, lead to a much broader whole. Turkey is situated in the first fringe, i.e. the one closest to the heart of the European region outlined, at the same level as Morocco or Azerbaijan, except that Turkey has been an official candidate for the EU since the 1990s, after several attempts (candidatures to the EEC in 1963 and then 1987): from the point of view of the intentionality of the actors and the history of its international relations, Turkey has long claimed to be European.

Such a representation remains strongly Eurocentric. Figure 11.2 shows that from the point of view of the regionally oriented IGOs to which it belongs, Turkey is not "between three worlds" (Bazin et al. 1998) but in five distinct dimensions. In addition to the European dimension already discussed, there is:

– A "Western" dimension, which takes shape through NATO and the OECD. The North Atlantic Treaty Organization is originally the European side of the system of alliances organized around the United States since the Cold War. It has expanded considerably since the fall of the Iron Curtain, integrating a whole group of states from the former Soviet glacis, two countries associated with the EU (Iceland and

Norway) and the two historical North American members (the United States and Canada). Historically, the Organization for Economic Co-operation and Development has been a socio-economic definition of the "West", grouping together countries that adhere to the principles of political and economic liberalism. It includes many of the United States' historical allies (from Australia to Chile to South Korea). Like NATO, it does not cover the European dimension, with a certain number of countries outside of it being members and others not (e.g. Malta, Cyprus, Georgia).

– A second dimension refers to Islam through the Organization for Islamic Cooperation (OIC). It comprises a vast group of states from the Maghreb to Southeast Asia, including parts of Central Asia and Sub-Saharan Africa. Very few countries share, as it does, a membership in this organization and in one of the European IGOs (Albania and Azerbaijan, for example, members of the EC). This is a way for Turkey to place itself on the map in this part of the world, even though it is not a member of the Arab League.

– A third dimension corresponds to a "Turkish world". Turkey has initiated organizations for cooperation between Turkish-speaking countries, such as the Turkish Council and the TÜRKSOY organization (international organization for Turkish culture). The latter includes among its members a number of Russian and Moldovan sub-state territories, as well as a state that only Turkey officially recognizes: the Turkish Republic of Cyprus (North).

– A final dimension of Turkish foreign policy is organized around the title "Black Sea" through the Organization for Black Sea Economic Cooperation, of which it has been a founding member since 1992. With the multiplication of conflicts between member states since the beginning of the 2000s (Russian–Georgian conflict of 2008, Ukrainian crisis of 2014), this structure is no longer active.

Thus, even when captured through official divisions of the boundaries of the territories covered by IGO member states, the geography of a European region is not straightforward and can be likened to a blurred whole (consisting of a core and fringes). Moreover, this region is not isolated, and as the case of Turkey suggests, some states may also fall within the scope of other dimensions of their neighborhood, making the space anisotropic and the identification of a particular region in isolation, such as Europe, of little relevance to understanding the situation of that country.

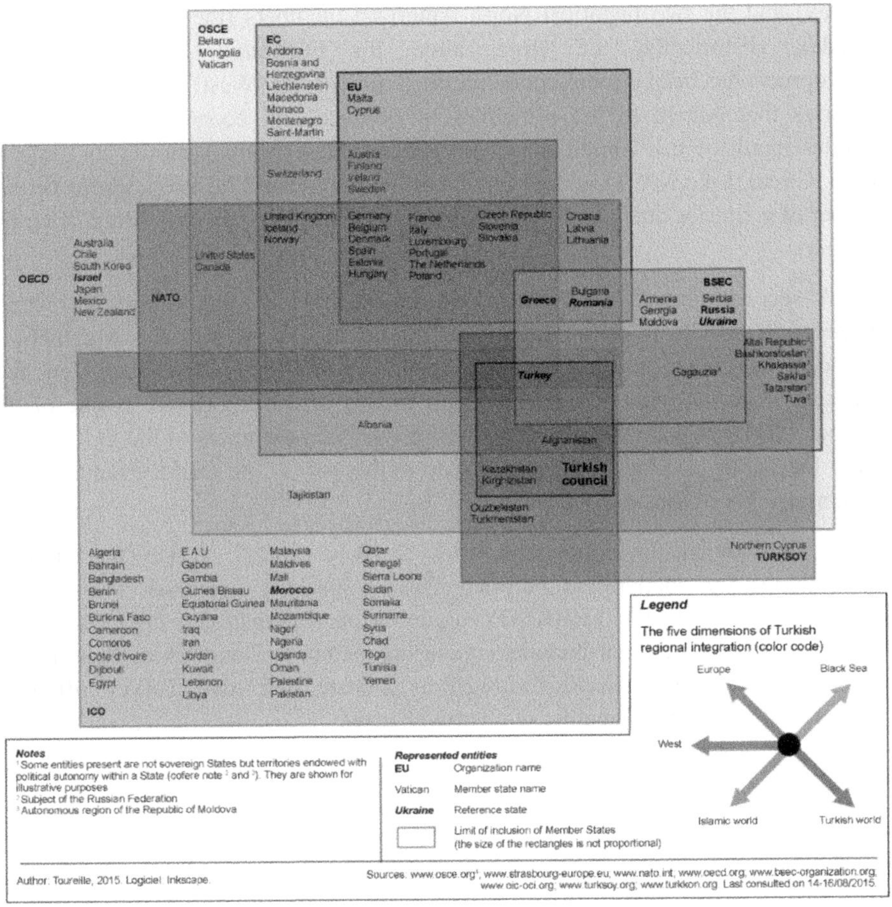

Figure 11.2. *The five dimensions of Turkish foreign policy based on its participation in IGOs. Source: Toureille 2017. For a color version of this figure, see www.iste.co.uk/beckouche/regionalization.zip*

11.4. Europe and world regions in the French national daily press (2013–2020)

A final approach to the European region analyzes its representations in the daily press, with the French context as an example. This involves studying the relative place of Europe in the regional division of the world through the analysis of a selection of place names (countries or regions of the world) mentioned in articles published in four daily newspapers of national scope: *La Croix, Le Monde, Le Figaro and Libération.*

The analysis covers more than 1 million news items published between July 2013 and June 2020 on the newspapers' RSS (Really Simple Syndication) feed. The data collected comes from the Media Cloud platform (see: https://mediacloud.org/). RSS is a technology that allows for the streaming and collection of information via a feed linked to an address. Thanks to an aggregator, it is possible to collect the information broadcasted by a media source on the feed during a given period. In the case of the daily newspapers studied here, the textual content of the RSS feeds includes the title of the article published on the newspaper's web page as well as a description, which we have limited here to the first three sentences.

To build the database, an automatic detection of geographical entities contained in the text of the news was carried out. In order to try to evaluate the level of belonging of the states to geographical objects of regional level such as "Europe", a textometric analysis of the copresence of states and regions in the same sentences was carried out. The statistical treatment aims to understand which are the most frequent or the rarest associations between two places, and to deduce a proximity both event-related (since it is press news) and editorial (since it is the journalists who choose the names of places used to report it). The analysis does not always distinguish these two forms of proximity. For example, Ukraine was frequently associated with Russia in 2014 because of the annexation of Ukraine (event proximity). But the fact that the term "Europe", "European Union" or "West" is used instead to describe the same event is often an editorial choice that can vary from one newspaper to another.

For example, let us consider the following news:

> *Germany* ranks first among EUROPEAN UNION countries for the number of asylum seekers, rising sharply with the influx of refugees fleeing war or violence in countries like *Syria*, *Iraq*, *Libya* or *Eritrea*, or poverty in AFRICA, the MIDDLE EAST or the BALKANS (*Le Figaro*, August 18, 2015).

In this sentence, five states (in italics) and four regions of the world (in small capitals) are detected, which are linked by the event reported. That is, a total of 10 relations between different countries (e.g. Germany–Syria) but also six relations between different regions of the world (e.g. European Union–Africa) and 20 relations between a country and a region of the world (e.g. Germany–European Union). Although each type of relationship is of interest in political geography, we will limit ourselves here to the last two cases, which are the most likely to reveal a political and regional geography of the world as seen by the French media.

Although the list of world regions used is not exhaustive, the results of the analysis confirm the major weight of the EU and Europe as a world region in media representations. Table 11.2 shows that out of a total of 43,681 detections of a world region or regional organization, there are 13,610 occurrences of the terms "European Union" (31%) and 12,246 occurrences of "Europe" (27.9%). The other regions of the world represent much lower weights: "Africa" (7.8%), "Asia" (5.1%), "Middle East" and "Near East" (3% and 1.2%), "America" (2.6%). This is obviously due to the fact that the media studied are French, and that news dealing with subjects directly related to France or its immediate vicinity are mechanically more numerous. As for regional organizations, we mainly note the presence of NATO (5.3%) and the African Union (1%). Finally, it should be noted that the seas and oceans, which we have chosen to add to the list because they also define a form of regionalization of the bordering countries, are present through the Mediterranean (2.3%), the Pacific Ocean (1.5%), the Atlantic Ocean (0.8%) and the China Sea (0.7%).

Rank	Region of the world	Workforce	Number per 1,000 occurrences of a region
1	European Union	13,610	310
2	Europe	12,246	279
3	Africa	3,426	78
4	NATO	2,343	53
5	Asia	2,259	52
6	Middle East	1,325	30
7	America	1,135	26
8	Mediterranean	1,025	23
9	Sahel	673	15
10	Pacific Ocean	659	15
11	Middle East	527	12
12	African Union	425	10
13	Atlantic Ocean	375	9
14	Amazon	290	7
15	S. China Sea	288	7
16	Caribbean	288	7
17	Balkans	281	6
18	Maghreb	239	5
19	Council of Europe	208	5
20	MERCOSUR	205	5

Table 11.2. *Most cited regions of the world in the press news of four French newspapers (2013–2020)*

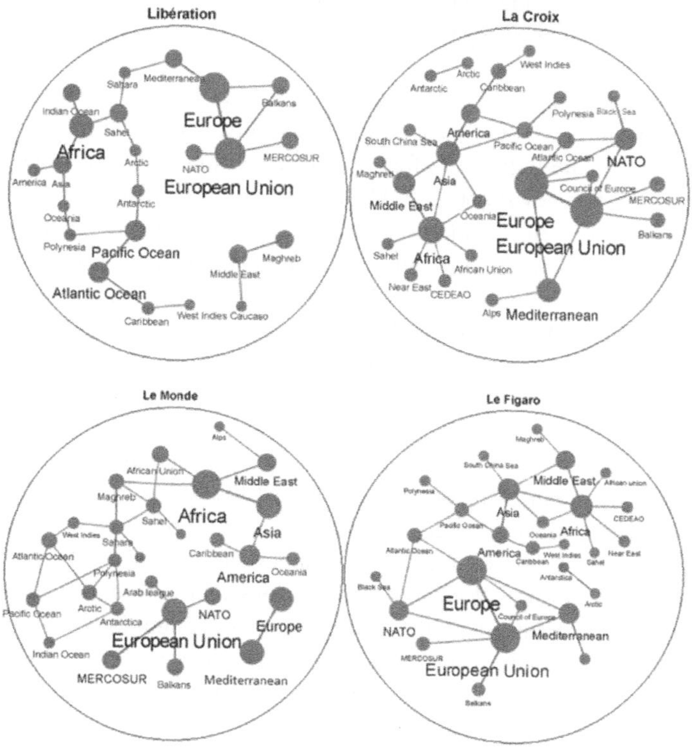

Figure 11.3. *Most significant world region associations in press news from four French newspapers (2013–2020).*

Given the overwhelming weight of the Europe/EU binomial, a simple representation of the most frequent associations between countries and regions of the world would mechanically lead to the formation of a double star linking each country of the world to these two entities, insofar as the probabilities of association for these terms are very much higher than those for the rarer terms. It is therefore more interesting to represent nontrivial associations, i.e. cases where some countries are significantly more often associated with one world region than with another, all things being equal as to their frequencies of occurrence[1].

1. To establish that a relationship is nontrivial, a random distribution model is constructed in which the number of links is a function only of the number of times the two places are mentioned. Then, we compare this theoretical number to the observed number and test the significance of the differences. We visualize only the relationships characterized by a positive excess of relationship with less than a 5% chance of being the effect of chance.

An initial analysis of the relationships between regions of the world was carried out for each individual newspaper (Figure 11.3).

The comparison of the graphs highlights strong similarities and reveals the presence of strong and significant connections between the terms "Europe" and "European Union": the terms are often both cited within the same news item, with the exception of the newspaper *Le Monde*. These two terms are then regularly associated with other neighboring regions of the world (Balkans, Mediterranean) or with alliance or membership systems of a political nature (NATO, Council of Europe). Some recurrent links can also be explained by significant events during the observation period. MERCOSUR, for example, appears to be the subject of recurrent negotiations with the EU between 2013 and 2019 with a view to establishing a free trade treaty.

The figure also highlights the regions of the world that play an important role in the French press without being so strongly linked to the EU/Europe binomial. The considerable weight of Africa and the Middle East, and to a lesser degree Asia and America, can be seen, confirming the still important role of the continents in the representation of the world. Moreover, the various maritime and oceanic areas as well as the poles are linked to the extent that they often form strongly related sub-graphs. This interest in distant maritime spaces is perhaps an indicator of the colonial heritage and media coverage of the overseas departments and territories.

A second analysis focuses specifically on the privileged relationships between state and regional mentions (Figure 11.4).

This figure differentiates countries that are associated by French newspapers with Europe, such as Germany or the Netherlands, while others are more associated with the EU, such as Canada or the UK. These are probably institutional issues related to the consequences of the Brexit. Several countries are in an intermediate situation between "Europe" and the "European Union", notably the countries that were at the heart of the debt crisis during the period (Greece, Ireland) or the countries of Central and Eastern Europe. Other cases of intermediate situations appear, notably for countries that are associated with both the EU and NATO (Russia, Ukraine, Turkey, Poland) or with the EU and the Mediterranean (Italy, Spain). If we broaden the analysis to the whole world, the EU and Europe are more closely linked to the United States, NATO, the Mediterranean and the Middle East. In the other parts of the world (Asia, Africa, the Americas), we note a set of classic – or at least expected – relationships in the texts of the corpus linking states to their classic continental region of reference.

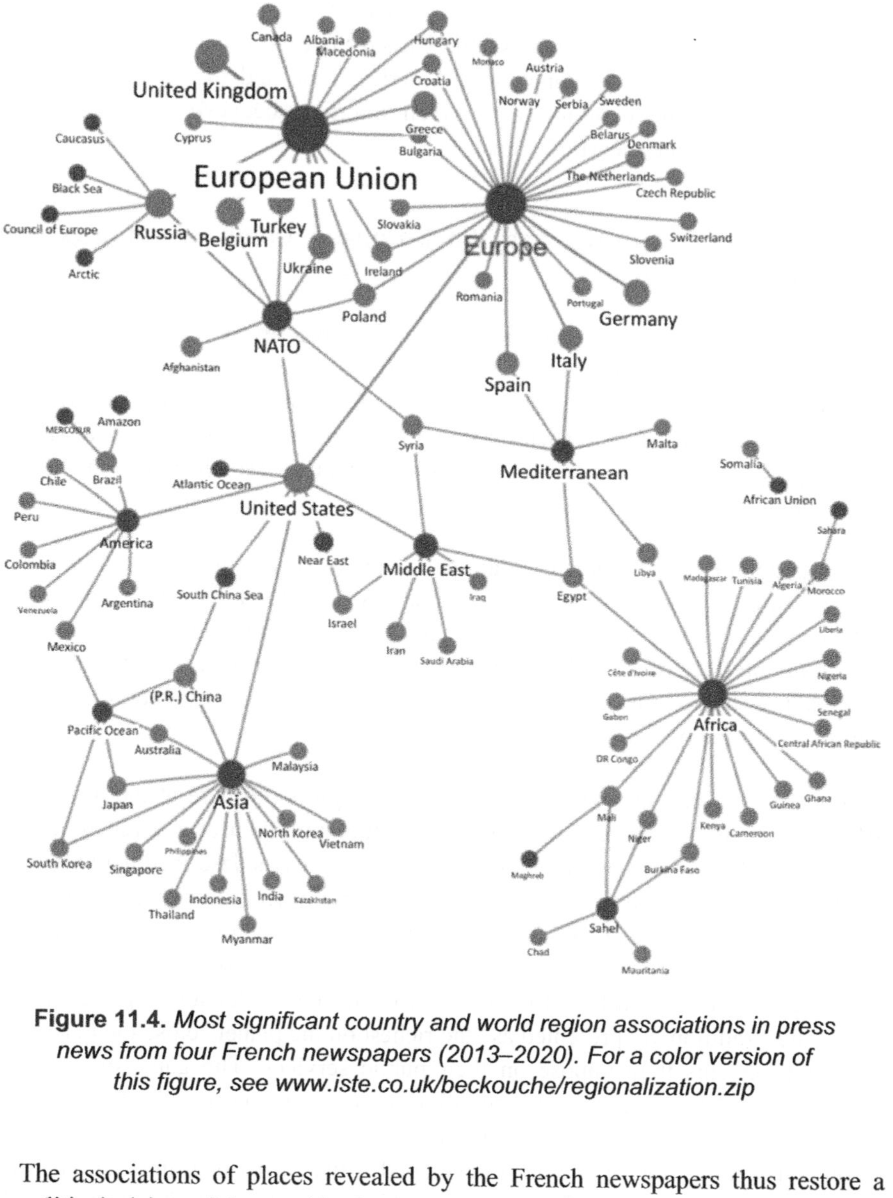

Figure 11.4. *Most significant country and world region associations in press news from four French newspapers (2013–2020). For a color version of this figure, see www.iste.co.uk/beckouche/regionalization.zip*

The associations of places revealed by the French newspapers thus restore a geopolitical vision of the world where Europe and the EU form a double star that constitutes an important anchor point in the description of events taking place in the vicinity, but neither more nor less than the other continents and other geographical objects (e.g. the Mediterranean) or political objects (NATO). In terms of centrality in the graph of media representations, the United States still occupies a privileged position since it links news concerning America, Asia and the EU–Europe. The Mediterranean, on the contrary, acts as an interface between news about the EU–Europe, Africa and the Middle East.

11.5. Conclusion: Europe and the European Union, a complex and ever-changing landscape

These three examples of analysis attempt to objectify a reality that we had presented, Europe, as volatile and polysemous. Defining Europe, for Europeans, inevitably leads to painful questions of identity, of possible decline and of a difficult balance between identification with states which, for many, taken in isolation, are average powers and supra-state ensembles in permanent search of legitimacy and coherence.

The geography of diplomatic relations invites us to lay the foundations of the spatiality of power in its materiality and practices. That of the inscription of Turkey in the complex game of the IGOs to give an example of multiple memberships of a state at the interface of several regional sets. The media, for their part, present a sort of sounding board for collective representations and confrontations between territories inspired by current events.

In all three cases, at first glance, the situation proposed here at the turn of the 2010s and 2020s would remain a kind of photograph taken at a given time. Our hypothesis would be that in its broad outlines, the processes described here impose a significant inertia, resulting for many from the world order recomposed after World War II and then after the end of the Cold War. But in reality, this is not incompatible with small and incessant changes, initiated by moments of crisis. Every year, emerging economies (Turkey, for example) open new diplomatic representations, while states traditionally presented as "the richest on the planet" close theirs because of the budget cuts they impose on their public services. The game of alliances, perceptible in IGOs, is constantly reconfiguring or changing the way they operate. Some are getting stronger, and others are becoming empty shells. The European Union is asserting itself as the most developed entity of regional integration, while having recently and for the first time in its history, lost one of its members. The media presents variable associations of places, influenced by specific events (Ukraine crisis of 2014, Syrian War, etc.). But the map of the world that they reveal finally remains quite close to the representations of common sense in the long term because it is anchored in school learning.

Nevertheless, from this shifting reality, from which it seems that only snapshots can be taken, it is possible to propose several interesting lessons. The first is that more than ever, the regions of the world are often defined by the sum of the states that make them up. This is true because of the institutional inclusion of these entities in supranational structures and therefore because of the logic of belonging to structures with the same view. And even in the economic field, which could be considered independent of national structures, especially in a system with relatively open borders, international trade is observed on the basis of territories that place companies in a particular legal and fiscal context. Thus, concerning the role of states and their relations with even very accomplished entities such as the EU, we can consider the glass as half empty or half full. On the one hand, the states are indeed decisive in their capacity to create discontinuities on their doorstep, thanks to their normative capacity. Moreover, they are also in a position to integrate themselves into groups or to withdraw from them according to their interests. On the other hand, the importance of regional integration in an abstract geographical entity ("Europe" or an international institution (the EU)) illustrates the inability of ultimately small states to deal with a certain number of contemporary issues such as global warming, energy production, migration and the maintenance of a successful agriculture. The claim to great power status is not a serious horizon for any European state, but regional integration may offer an opportunity to be visible and an unprecedented capacity for action on a global scale, which is nevertheless in the making.

11.6. References

Bachmann, V. (2022). Visionary geographies and European studies. *Progress in Human Geography*, 46(2), 416–440.

Badie, B. (2019). *L'Hégémonie contestée*. Odile Jacob, Paris

Bazin, M., Kançal, S., Pérez, R., Thobie, J. (1998). *La Turquie entre trois mondes : actes du colloque international de Montpellier, 5, 6 et 7 octobre 1995*. Institut Français des Études Anatoliennes (IFEA), Istanbul.

Beauguitte, L. and Didelon, C. (2011). Structures spatiales des relations diplomatiques mondiales. *Dixièmes rencontres de Theoquant*, 10, 1–10.

Brams, S.J. (1966). Transaction flows in the international system. *American Political Science Review*, 60(4), 880–898.

Neumayer, E. (2008). Distance, power and ideology: Diplomatic representation in a world of nation-states. *Area*, 40(2), 228–236.

Toureille, E. (2017). La Turquie en Europe, la Turquie dans le monde : analyse d'une catégorie macro-régionale au regard d'une géographie des représentations de l'Europe et du monde turco-centrée. PhD Thesis, Sorbonne Paris Cité.

Xierali, I.M. and Liu, L. (2006). Explaining foreign diplomatic presence in the U.S. with spatial models: A liberal spatial perspective. *GeoJournal*, 67(1), 85–101.

12

The Arctic, a (Macro)Region under Construction?

Camille Escudé[1] **and Frédéric Lasserre**[2]

[1] *CERI-Sciences Po, Paris, France*

[2] *Université Laval, Quebec, Canada*

12.1. Introduction

Perhaps more than for other regions of the world, there is no consensus on the definition of the limits of the Arctic. Several definitions coexist and overlap, depending on the disciplines and points of view: latitude 66° 33' north, which defines the Arctic Circle; the 10° Celsius isotherm, which connects all points where air temperatures remain below ten degrees during July; the tree line, which marks the southern continental limit of treeless vegetation typical of the polar climate (tundra); the limit of the annual mean sea ice extent or the Arctic polar front, which marks the separation between the cold, low-salt "Arctic waters" and the relatively warmer, higher-salt waters. These definitions are relevant to a given specialty (ecology, oceanography, climatology, etc.) and subject to variability in the context of climate change. If these definitions are devoid of any cultural and social dimension, none of them can claim to be universal.

In addition to these scientific definitions, each nation has its own definitions. In Canada, the administrative definition of the southern limit of the three northern territories (Nunavut, Northwest Territories and Yukon) is latitude 60° north. In the Nordic countries, the Norwegian North (Nordland, Troms and Finnmark), the Finnish North (Laapi, Kainuu and Northern Ostrobothnia) and the Swedish North

Regionalization of the World,
coordinated by Pierre Beckouche and Yann Richard.

(Norrbotten and Västerbotten) have a southern boundary near the Arctic Circle. In 2013, the Russian Arctic zone was administratively redefined to identify priority areas for economic development under the Russian Federation's Arctic 2020 strategy. Although Iceland is located just below the Arctic Circle, its Arctic strategy claims it as an Arctic Ocean coastal state, arguing that from an economic, political and security perspective, the Arctic extends from the North Pole to the northern Atlantic Ocean. Other Arctic limits are purely military, such as the Dew Line (Distant Early Warning Line), consisting of an alignment of US and Canadian radar bases on the 69th parallel. This line, operational since 1957, linked bases located in the Canadian Far North, on the north coast of Alaska, in the Aleutian Islands, in the Faroe Islands, in Greenland and in Iceland, in order to detect any attempt of Soviet intrusion during the Cold War.

Finally, emphasizing that the North is both imaginary and physical, the Quebec geographer Hamelin (1975) was the first to explain that the very identity of the Arctic seems to be reduced to its establishment around the imaginary line of the 66th parallel, whereas other limits are more appropriate. In the 1960s, he proposed the notion of nordicity, which is measured from biophysical, and also human and socio-economic elements, and is completely new for a region that until then was only considered for its physical characteristics. Nordicity is a tool that shows the changing character of the various Norths, since some territories are experiencing a "denordification" due to increased accessibility, the appearance of new cities or climate change. On the contrary, the identification of "pre-North" zones makes it possible to break away from the idea that the Arctic would be circumscribed by a fixed limit (Escudé-Joffres 2020a).

There are therefore several Arctic regions, with varying geographical extensions and framings. Keep in mind that the possibility of shifting boundaries is necessary to understand this region as an object in motion, in the process of definition. Knowing that the creation of regions is conditioned by politics, cultures, economies, governance and power relations in a contingent manner (Paasi 2010), how can we then define the Arctic region, a fragmented and dynamic space, moving with political and environmental changes?

First, we will show why and how the notion of region is mobilized by the Arctic actors. Then, we will see that the Arctic region is above all constructed from above, through a process of institutional regionalization carried out for political purposes. Finally, we will question the construction of a functional Arctic region, which is still in its infancy at the macro-regional level, but which becomes clearer as we increase the scale of observation.

12.2. The Arctic, a self-proclaimed region? The notion of region mobilized (or not) by the Arctic actors

12.2.1. *Belonging to the Arctic, previously politically mobilized*

The desire to belong to the Arctic region seems to be politically motivated. While the eight states with territories north of the Arctic Circle (Canada, Denmark, the United States, Finland, Iceland, Sweden, Norway, Russia) claim to belong to a common region, they are at the same time reaffirming their sovereignty over these territories in a general context marked by a growing interest in this part of the world (Pic 2018). We can recall, for example, President Donald Trump's proposal to buy Greenland in August 2019, a way for the United States to reaffirm its presence in the Arctic vis-à-vis Russia, perceived as a new threat and vis-à-vis China (Escudé-Joffres 2020a). In the declarations of the Arctic Council, composed by the above-mentioned states, their belonging to the Arctic is regularly reaffirmed. Since they signed the Ottawa Declaration that founded the Arctic Council (Arctic Council 1996), these states have presented themselves as the only "Arctic States". Moreover, countries applying for observer status in the Council must first recognize the sovereignty of the Arctic States. They must prove that they respect the values of the populations of the Arctic States as well as their "interests" in the Arctic. This requirement is explained by the fear of a Chinese candidacy. This criterion is deliberately vague. It therefore gives the Council's member states considerable leeway to reject the application of any state (Arctic Council 2013), like many regional organizations that restrict the admission of new member states.

Attempts by some states to interfere in Arctic affairs and challenge the sovereignty of the states in this region have been very badly received, in particular the Proposal by the European Parliament of an Arctic Treaty based on the Antarctic Treaty in 2008. In its resolution, the European Parliament asked the five countries bordering the Arctic (Canada, Denmark, the United States, Norway and Russia) to renounce their rights under the Montego Bay Convention on the Law of the Sea, in order to strengthen the governance of the region by signing a treaty similar to that of the Antarctic continent (European Parliament 2008). We can also think of China, which, in its white paper, presents itself as a "near-Arctic State" (The State Council – People's Republic of China 2018), which has elicited mixed reactions, even to the point of American indignation (Escudé-Joffres 2020b), with the American Secretary of State Mike Pompeo repeating in 2019 that there are only Arctic and non-Arctic States, and nothing in between. Nevertheless, it can be noted that the United

Kingdom has developed a similar concept[1] (Polar Regions Department 2018), without incurring the wrath of Washington.

12.2.2. *Other minimal forms of belonging: economic, autonomous, scientific*

While there are many speeches affirming that they belong to a common regional grouping, some actors – admittedly fewer in number – claim to belong to a common economic region. The creation of the *Arctic Economic Council*, whose stated aim is to develop economic and commercial cooperation in the Arctic, is an illustration of this. This Council is an independent entity where "*business representatives*" appointed by each member state of the Arctic Council sit. The Council's mission is to promote sustainable economic development in the Arctic by facilitating economic activities and investments, particularly for the benefit of indigenous peoples and small businesses (Gattolin 2014). Proclaiming to belong to an Arctic region is a territorial marketing tool used by some territories, notably Lapland where polar tourism is growing very rapidly (Hall 2009). In this region, there has been a doubling of the annual number of overnight stays booked by foreign tourists between 2010 and 2019, to over 3 million tourists in 2019 (Lapland.fi 2020).

On the contrary, it is the cultural dimension of belonging to the Arctic that is emphasized by the indigenous populations, which represent 15–20% of the region's inhabitants (1CIA 2004). These populations are represented on the Arctic Council by six organizations. There are also transnational indigenous organizations, such as the Sami Council, which brings together groups residing in Finland, Norway, Russia and Sweden; or the Inuit Circumpolar Council, which claims to represent 180,000 Inuit residing in Alaska, Canada, Greenland and Russia (Chukotka Peninsula). However, no formal organization carries or prefigures any kind of regional political integration of the indigenous peoples of the Arctic.

Another form of integration is that of epistemic communities, i.e. professional networks that are distinguished by the possession of certain skills, know-how or expertise applied to specific fields. Researchers can form such a community. Moreover, the production of scientific knowledge on and by the Arctic territories has long been a powerful vector of political cooperation, which is at the origin of the creation of the Arctic Council (Escudé 2019). Scientists form transnational working groups and aggregate their research. This research leads to policy recommendations and provides knowledge that is the primary material for the work of the Arctic

1. "Although the UK is not an Arctic State, we are its nearest neighbour" (Polar Regions Department, 2018, p. 2).

Council. In doing so, the transnational scientific working groups create cooperative links between actors in the region. International experts share best practices, norms, values and interests; they work together and thus participate in a regional scientific network. It is difficult to measure the effectiveness and scope of scientific work in the field, in other words, how it is considered in the policies implemented by the states. However, the socialization between actors induced by the dynamics of the epistemic communities is real and the sharing of research results is effective through the publication of voluminous research reports by the working groups of the Arctic Council (Landriault et al. 2020)[2]. The meetings of researchers in different Arctic countries, the networking in the working groups and the joint publications in scientific journals make the Council a real place for the development of social practices that bring the actors together. It can thus be said that the circumpolar dimension of science has played a founding role in the creation of an Arctic governance.

In contrast, any military dimension is rejected out of hand in the Arctic, even though political affiliation is the main condition for regional integration in this part of the world. The refusal to discuss security issues multilaterally is based on a consensus among states following the Cold War, which led them to focus on civilian rather than military issues in their political cooperation. Peace and security issues, central issues in this former Cold War "hot spot", are completely excluded from the Arctic Council's purview, which many observers believe contributes to its success (Escudé 2017).

The fact that actors claim to be Arctic in different fields suggests that social and political practices are a first way of constructing the region, including in discourses. The Arctic region is above all constructed by national public institutions. It is institutional regionalism, carried by the practices of governments and public institutions, that draws the boundaries of the region. However, these limits are sometimes contested because they are precisely constructed from above.

12.3. A region built mainly from above: institutional regionalism draws the boundaries of a contested region

12.3.1. *Protecting the environment: a pillar of regional integration*

In the Arctic, regionalism is driven by public institutions. It was really launched in 1987 with the speech made by Mikhail Gorbachev in Murmansk. The Soviet

2. The working groups are: Arctic Contaminants Program; Arctic Monitoring and Assessment Program; Conservation of Arctic Flora and Fauna; Emergency Prevention, Preparedness and Response; Protection of the Arctic Marine Environment; Sustainable Development Working Group.

leader proposed to develop Arctic political cooperation through circumpolar partnerships in several fields (environment, science, economy). The motivation was partly political. Indeed, the cooperations that multiplied afterwards (e.g. the Arctic Council, the Council of the Baltic Sea States, the Committee of Parliamentarians of the Arctic Region, the Barents Euro-Arctic Council, etc.) were probably aimed at enabling Russia to participate in Arctic political cooperation (Browning 2003). The other common objective of these cooperations was and still is to promote environmental protection. This has become a sort of cornerstone of functional cooperation and the pretext for meetings between all the Arctic actors. The protection of the environment and the promotion of sustainable development are indeed the two objectives of the Arctic Council (1996), recognized as the privileged and central organization of cooperation in the region, where all the Arctic States discuss regional issues, leaving aside their diplomatic or political differences, as was the case during the Crimean crisis. This model of cooperation is flexible and non-binding, and it facilitates relations between diverse actors with a priori opposed interests (Escudé 2017), ultimately making science a pretext for political cooperation. This situation is seen by some observers as an "Arctic exceptionalism" (Exner-Pirot and Murray 2017; Käpylä and Mikkola 2019).

The cooperation forums in the Arctic participate in the construction and organization of this region. If it has been proven that, in the case of Europe, political cooperation can lead to regional integration (Mattli 1999), a similar process can be found in the Arctic. Indeed, the protection of the environment and the promotion of sustainable development play a leveraging role in the regionalization process at work. Regional integration involves the creation of institutions such as the Arctic Council, whose mission is to achieve the objectives of integration and maintain them over time (Girault 2018; Escudé-Joffres 2021), by providing actors with a framework for formal and informal relations. Political cooperation then contributes to strengthening regional integration, and regional integration increases as the actors who benefit from it strengthen their cooperation. In this virtuous circle, the sharing of certain constraints (distance and remoteness, the presence of pack ice, extreme cold, the vulnerability of ecosystems, etc.) and the interactions that exist between actors at various levels involved in managing the risks related to these constraints have fostered a form of regional political cohesion. Since the Murmansk speech, the leaders of the eight Arctic States have met regularly and maintained diplomatic cooperation in the framework of the Arctic Council and its working groups. Regional institution building allows the Arctic states to consolidate their influence in the region and ensure that their interests are considered by non-Arctic countries, which are increasingly challenging this political framework.

12.3.2. *External actors attempt to expand the political boundaries of the region*

Since the 2000s, the Arctic has been of growing interest in various countries, in relation to the development possibilities of extractive industries and transportation (Lasserre et al. 2021). This interest, mainly scientific since the 19th century, took on a more economic and financial dimension as the representation – probably overrated – of a region with strong strategic and economic potential was constructed (Lasserre 2021a, 2021b), and then a political dimension. When the Arctic Council was created, four observer states were present (Germany, the Netherlands, Poland and the United Kingdom), but the number of countries applying for this status has not stopped increasing since then. In 2013, five new observer countries joined the Council (South Korea, China, Italy, Japan and India). Since the entry of China and India, more than half of the world's population is represented on the Arctic Council. Even though the role of observers is very limited in the Council, this diplomatic status is highly sought after and allows them to obtain a significant symbolic place in the political arena of this part of the world. But, on the contrary, these external actors come up against the desire of Arctic actors to exclude them from regional decisions, on the pretext that some observers are indeed hardly involved in the activities of the Council's working groups and seem to be content with being present in the discussion forums (Babin and Lasserre 2019; Ren 2019) or harbor expectations that are sometimes excessive, even though they are not states of the region. For example, some states have requested observer status – hoping to gain more leeway in discussions and more weight in Council decisions – and have been disappointed, as observer status in a body that operates by consensus (consensus of member states and indigenous organizations) effectively gives little power (Rottem 2020). Through the voice of the Ambassador in charge of negotiations on the poles – Michel Rocard (between 2009 and 2016) – France has thus criticized a management of the Arctic that is considered too closed. It has repeatedly criticized this so-called club diplomacy, believing that the Arctic was "managed like a condominium" by the riparian states (Lasserre et al. 2021). Based on the model of the Antarctic Treaty, of which he was one of the instigators, Michel Rocard proposed to include states outside the limit of the Arctic Circle in the governance of the zone to promote its sanctuary. This proposal was not well received by the sovereign states in the Arctic, who argued that the Convention on the Law of the Sea applied in the region. This illustrates the tensions over the boundaries of the decision-making perimeter for the region. Many observers of the Arctic Council believe that the current limited governance model – including only the eight member states and representatives of indigenous peoples – is reaching its limits and are questioning it. On the contrary, member states and indigenous organizations argue that it is legitimate for regional affairs to be managed by local actors, as in most regional organizations. These observers have been imaginative in bypassing existing political institutions or inventing new forms of political action to make their voices heard in alternative forums. Following initiatives by states outside the Arctic, by governments of federated political entities such as Scotland, and by NGOs, the boundaries of the Arctic region are becoming blurred and are increasingly shifting to places and spaces outside the boundaries of the Arctic Circle (Escudé-Joffres 2020b).

The use of regional institutional arrangements: bypassing traditional bodies and creating new forms of political action

International Arctic conferences are currently privileged platforms for the expression of actors usually excluded from formal cooperation in this part of the world, as evidenced by the international conference "*Arctic Circle*" held in October since 2013 in Reykjavik. Former Icelandic President Ólafur Ragnar Grímsson established the conference in 2013 to form "the largest network for international dialogue and cooperation on the future of the Arctic" (Arctic Circle n.d.). For companies, non-Arctic or even non-riparian states, or for sub-national governments such as Quebec or Scotland, international conferences allow for more open cooperation involving a larger number of actors. These encounters take place through meetings, small committee sessions or more informal gatherings. The "*Arctic Circle*", a self-proclaimed "democratic and open platform", is tacitly opposed to formal organizations that are considered closed, such as the Arctic Council. It presents itself as "the largest international gathering on the Arctic" and, notably, the topic of discussion is "Arctic development" but not its protection. It is also possible to see a fashion effect in this infatuation with the supposed wealth of the Arctic, as if we have to be part of a regional governance process to benefit from it. Several cities, countries or federal governments have hosted a delocalized forum. On the state side, we note China, South Korea and Singapore; on the non-state side, the Faroe Islands, Greenland, Scotland and Quebec. Thus, there is both a significant eccentricity in geographical terms towards Asia, and a political eccentricity towards non-state spaces. The "*Arctic Circle*" conference is even considered by some actors as a relevant alternative political format for those excluded from the Arctic Council (Escudé-Joffres 2021). The majority of actors with a strong presence at this conference have the fact that they are considered "outsiders" in Arctic politics in common. For Iceland, its leadership in creating a conference is explained by its somewhat marginalized position in the Arctic and its search for legitimacy as a littoral state. Federated states and non-member states of the Arctic Council attend these conferences alongside companies, think tanks, various pressure groups and researchers with little political weight in the official bodies of a regional governance that is increasingly closing in response to this international pressure (Escudé-Joffres 2020).

All informal and so-called open forums offer a chance for non-traditional actors, i.e. non-circumpolar, non-state, or both, to access Arctic political networks. This illustrates the broadening of Arctic regionalization, geographically with its concentric circles extension, and politically with the inclusion of non-state regional actors (Escudé-Joffres 2021). By focusing our attention on the sub-state level with the Scottish example, we observe that these actors are keen to move the boundaries of the Arctic region in order to claim their own geographical belonging, which then becomes a lever for including or excluding certain actors from political decision-making procedures.

The involvement of sub-national governments in Arctic governance: a political paradiplomacy that expands the boundaries of the region

These assemblies are a privileged place for sub-state administrations to express themselves and to lobby, and they provide a political forum for actors such as Scotland or Quebec, who are often excluded from official bodies. The concept of paradiplomacy emerged in the 1980s to describe these new forms of diplomacy, as part of the revival of federalism and comparative politics studies. This notion refers first to the international activities of Canadian and American-federated states in the context of globalization (Paquin 2004b), and, in a broader sense, to the international activities of sub-state entities in general. Paquin proposes a definition of identity-based paradiplomacy as opposed to traditional paradiplomacy. With regard to the latter, the objective of what he calls "identity entrepreneurs" is to obtain political independence. The goal of identity-based paradiplomacy is to seek, at the international level, the symbolic and material resources that are lacking in order to be recognized as a nation at the global level (Paquin 2004a). This process of international legitimization is essential to any attempt to build a minority nation within the framework of a historically plurinational state.

Scotland, as a constituent nation of the United Kingdom, has been developing an Arctic paradiplomacy for several years, which is accompanied by an increased involvement in the political bodies dedicated to the region. It claims to be a full-fledged Arctic player, highlighting the British government's apparent lack of interest in the Arctic. This disinterest is interpreted by the Scottish authorities as an invitation to occupy a place left vacant by London (Johnstone 2012). First, it is to highlight Scotland's geographical proximity to the Arctic. The Shetland Islands, to the north of Scotland, are closer to the Arctic Circle than to the British capital:

Nicola Sturgeon, Scotland's First Minister, often cites this geographic proximity as a reason for a "special relationship" between Scotland and the Arctic (Lottero 2018). It is then the historical ties with the Arctic that are recalled in the speeches of Scottish political actors. Scotland thus believes it has a historical and cultural link with Iceland and Denmark, a link that predates the unity of Great Britain (Lottero 2018). Finally, the third argument invoked is that of economic interests that are substantially like those of northern Europe, in several areas: the extraction of hydrocarbons, fishing and the hypothetical opening of new trade routes.

Armed with these justifications, Scotland has been implementing a very active Arctic paradiplomacy for the past half-dozen years. The Scottish government hosted an "Arctic Circle" forum in Edinburgh in 2017 with the theme "Scotland and the New North". On that occasion, it released an Arctic Strategy (published in 2019), separate from the UK's strategy that was published in 2013 and then updated in 2018. At 46 pages long – more than twice as long as the UK's 2018 strategy – the "Arctic Connections" strategy published by the Scottish Government highlights Scotland's cultural, social and environmental links with the Arctic and concludes with "Scotland's Offer to the Arctic": i.e. concrete proposals for work between the two entities (Scottish Government 2019). It is not insignificant that advocates for an Arctic policy in Edinburgh are, with few exceptions, members of the Scottish National party (Depledge 2018). In the context of Brexit, some Scottish politicians are emphasizing their nation's northern identity to claim political – and geographic – distance from London.

This diplomatic process attests to a broadening of the political framework of the Arctic towards the south, as well as its political broadening because it includes actors that do not have the status of a state such as Scotland. The Arctic then consists of a combination of networks of contingent social relations. The Arctic is therefore not a region to be taken at face value. It can evolve and expand, or on the contrary disappear as a result of a process of "regional disintegration" (Richard 2018). In this regard, we can ask whether the suspension of circumpolar political cooperation caused by the war in Ukraine in 2022 prefigures a possible regional disintegration. *Conversely*, the Arctic region could experience an evolution caused by the possible integration of new actors that are not currently considered Arctic, but claim this status.

12.4. Building a functional Arctic region from below?

Despite the top-down regionalization of the Arctic for more than 40 years, the construction of a functional Arctic region is still to come.

12.4.1. *The weak regionalization of activities*

Contrary to Stefansson's intuition, who developed the idea of an "Arctic Mediterranean" in 1921, the melting of the ice pack does not lead to an increase in trade between the countries bordering it. The products of the Arctic countries are largely exported to nearby markets, but located further south and, increasingly, to distant third countries: deliveries of Siberian gas and oil to Japan and China, exports of iron ore from the Mary River mine to European countries and China. Conversely, there is little trade or passenger traffic between Arctic countries.

Between 2012 and 2015, the Iqaluit (Nunavut) airport hosted two flights per semester by Air Greenland from Nuuk, Greenland. These flights were discontinued as there were not enough customers to make the operation of this route profitable (Bell 2019). However, Air Greenland continues to operate flights to Iceland. For its part, Air Iceland Connect operates a flight between the Faroe Islands and Keflavik, and another one between Reykjavik and Greenland, thus providing an interconnection between Arctic and sub-Arctic regions. But this solution is not perfect because it requires switching from one airport to another in Iceland. In addition to the logic of hubs for long-distance connections between states, there are short-distance air services by cabotage, for example, in Norway and Canada: this makes it possible to serve numerous localities successively without crossing borders. Icelandair, for its part, has operated a transatlantic hub for several years and offers flights between North America and Europe (especially the Scandinavian countries[3]), with quick connections at its main airport in Keflavik on the west coast of Iceland. In doing so, it meets a commercial demand, but it does not constitute a lever for regional integration. Indeed, none of the airlines in this fast connection system serve Arctic regions.

Similarly, maritime connections are very few and compartmentalized state by state. Greenland has a coastal shipping service to the communities (for passenger transport), but no passenger shipping line exists across Baffin Bay, linking Greenland to Nunavut, despite only 890 km between Nuuk and Iqaluit. For cargo, the Canadian Arctic is served by the Canadian companies Woodward (based in St John's), Desgagnés and NEAS (Montreal), and MTS (Hay River). For its part, Greenland is supplied with consumer goods mainly by the Danish company Royal Arctic Line. This is part of an initiative by Royal Arctic Line and the Icelandic company Eimskip to cooperate in improving the supply of transport between Denmark, Iceland and Greenland (Humpert 2020a).

3. Flights to Bergen (Norway), Oslo, Stockholm and Helsinki.

The Arctic Bridge is an experimental intra-Arctic Sea route between Murmansk, Russia and the port of Churchill, located in Canada on the shores of Hudson Bay in northern Manitoba. Both ports are served by a railroad and have extensive hinterlands. The latter, opened in 1931 to export Canadian wheat to Europe, saw its traffic grow to 710,000 tonnes in 1977, then decline due to competition from exports to Asia. The port was privatized in 1997 and the operator OmniTrax then tried to diversify its activities, in partnership with the Murmansk Shipping Company, by shipping grain to Russia – via Murmansk – and by receiving chemicals imported from Russia, for example, fertilizers shipped from Murmansk (Friesen 2007; Lasserre 2010). But these attempts were not successful. After the dismantling of the Canadian Wheat Board in 2012 and due to a decline in traffic, OmniTrax decided to close the port in 2016. It was reopened in 2019, but its traffic remains fragile. Moreover, transportation between Churchill and Murmansk is no longer offered by the new operator Arctic Gateway (Arctic Gateway 2020).

The Northern Sea Route is widely operated from or to Russian ports. In 2019, 2,694 voyages were recorded on at least part of the maritime corridor, but only 37 voyages were transits between the Bering Strait and the Kara Strait, many of them to Murmansk. A few trips involved the transhipment of LNG in the Honningsvåg Fjord in Norway by the Russian company Novatek, but this was only a technical operation involving the loading of gas onto conventional LNG carriers from reinforced hulls. Thus, it was not a true service to the Norwegian Arctic (Staalesen 2019). There is, however, a regional infrastructure project to capture some of the possible container traffic that would pass through the Northern Sea Route (where there is no traffic at present) via a major port terminal located in Kirkenes (Norway). This port would then be served by a railroad linking it to Rovaniemi in Finland, which would allow it to be connected to the Finnish and then European rail network via a tunnel dug between Helsinki and Tallin in Estonia. But this very expensive project, costing more than 3 billion euros (Arctic Corridor 2019), is facing determined opposition from the Sami (Nilsen 2020).

In the railway field, only the European Arctic and Siberia have seen a significant development of the network, in connection with the exploitation of natural resources in Scandinavia (iron ore) and Siberia (Lasserre and Têtu 2020). In North America, the railroad reaches the city of Fairbanks in the heart of Alaska and Hay River on the southern shore of Great Slave Lake in Canada, but interconnection projects launched as early as 2000 between Canada and Alaska could not succeed because the investment required was too high and the size of the target market too small (Metz and Taylor 2012). A modified version of this project was given the green light by the US administration in October 2020 (Humpert 2020b). However, various analysts believe that this 2,570 km long rail line will not be built because it would cost US$17 billion.

The weak interconnection of Arctic spaces is not a surprise. The Arctic space is still characterized by very large distances. The circulation is still hindered at sea by the ice pack and more and more by the melting of the permafrost on land. Population densities are very low. Finally, economic activities are generally similar, and therefore not complementary: the Arctic regions mainly produce resources that are not consumed locally (ores, hydrocarbons, fish), but are shipped to markets further south along highly compartmentalized logistics chains, state by state. This makes it difficult to get intraregional trade off the ground or to offer profitable and sustainable transport services. Attempts have often failed. The only part of the Arctic world where greater interconnection is possible is around the Barents Sea. This explains why the construction of an Arctic region is driven by public institutions – notably through the Arctic Council – despite the economic fragmentation and a colonial-style operating model described above.

12.4.2. *Regionalization of trade on a larger scale: the example of the Barents region*

A certain regional functional cohesion emerges between Norway, Sweden and Finland (and even Russia), as the settlement, urban fabric and road networks reach fairly high levels of density. The road network serves the entire area, where there are many small communities and urban areas, often of modest size (Murmansk 287,000 h in 2019; Nikel 11,200 h; Hammerfest 8,000 h; Alta 19,000 h; Tromsø 71,000 h; Kiruna 22,900 h); there are also ice-free ports all year round and economic activity based mainly on resource extraction, as well as an economic fabric formed by many SMEs, which offer commercial services in particular. The population density higher than the average in the rest of the Arctic and the existence of adequate transportation networks – admittedly covering small distances but allowing for frequent border crossings – facilitate the movement and regional economic integration in the western part of the Barents Sea region. For example, the economic border between Norway and Russia is valued in the Kirkenes region: many Russians who reside near the border come to shop, allowing this small (3,200 inhabitant) settlement to have quite respectable sized shopping centers (Lasserre 2018). However, after the outbreak of the war in Ukraine in 2022, Norway and Finland imposed severe restrictions to border crossings for trade and port calls by Russian vessels. The Kimek shipyard in Kirkenes used to derive 70% of its revenue from fitting Russian ships; it is thus very vulnerable to a possible collapse in this activity (Fouche and Klesty 2022), and Russian consumers coming to shop in Kirkenes will have to adapt to the end of transborder car insurance agreements (Staalesen 2023), forcing them to adapt with bus services (Salama 2023). Murmansk has ended its relationship with its twin city Akureyri (Nilsen 2022). The impacts of the war on local and regional cross-border relations may not have ended transborder relations, but they have heavily impeded them.

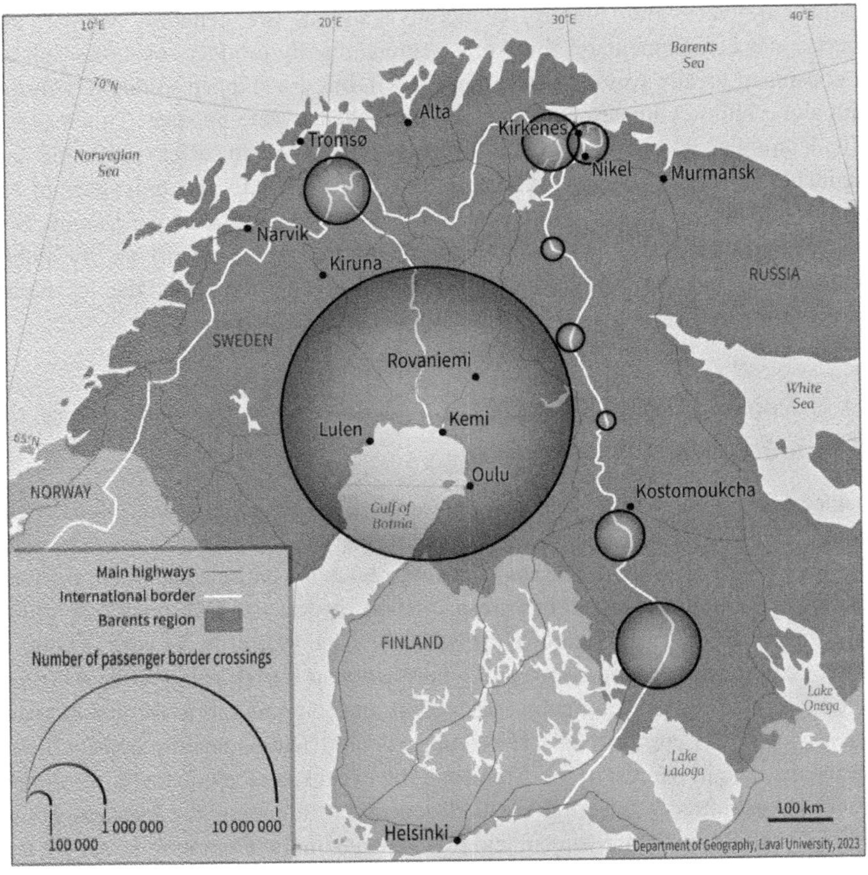

Figure 12.1. *Number of cross-border passengers, 2019 (source: http://www.barentsinfo.org/barents-region/Statistics/border-crossings)*

However, regional air services are still largely compartmentalized by state. In fact, SAS (Scandinavian Airline System, a Norwegian, Swedish and Danish airline with freedom of flight throughout the Scandinavian market) and the Norwegian, Swedish, Finnish and Russian regional airlines do not serve any of the border locations, even from the Norwegian hub in Tromsøu (Flight Connections 2020). However, the service is well organized in this part of the Arctic, with numerous and cheap flights[4].

Several initiatives have been taken to create regional routes. The Arctic Airlink Tromsø (Norway) – Lulea (Sweden) – Oulu (Finland) service, opened in January 2015 by the Swedish company NextJet, was closed in May 2018: only 9,700 passengers were carried and the average load factor on the aircraft was only 20%, i.e. an unsustainable load factor. In 1993, a Tromsø–Murmansk route was opened, with weekly flights offered by Braathens and Aeroflot. Braathens decided to close the route in 2000. Russian airlines continued operations for a while. But the last operating airline (Pskovavia) closed the route in 2014. The company Widerøe Flyveselskap opened a route between Kirkenes and Murmansk in 1994. This service was discontinued in 1998 by Aeroflot and in 2000 by Widerøe (Gustad 2001; Barents Euro-Arctic Region 2019). Air transport, which is not profitable, is not a lever for regional integration here.

Functional integration is both a criterion and an important sign of regional space integration, which has long been emphasized by geographers (Juillard 1962; Seys 2019). However, in this respect, we can only observe the weak internal integration of the Arctic zone, apart from its European part. There is little transport and little trade. This region remains structured by meridian flows towards the south of the Arctic states or towards more distant markets.

12.5. Conclusion

In the growing interaction between the local, national, macro-regional and global levels, the Arctic macro-region is playing an increasingly important role from a political perspective. In this case, regional integration can be understood as the contribution of Arctic institutions to the densification of institutional and functional connections among the inhabitants of the region (Stokke and Hønneland 2007).

Multidimensional, regional integration may be more advanced in some areas than in others (Richard 2018). The Arctic, as a geographic region, is involved in various cooperation mechanisms that tend towards a pooling of political means.

---

4. In Canada, a round-trip flight from Montreal to Iqaluit cost US$1,920 in October 2019 on First Air (before its merger with Canadian North in late 2019), more than a transatlantic flight; a return flight from Oslo-Kirkenes, US$340 on SAS and Tromsø-Hammerfest, US$185 on Wideroe. In November 2021, the prices were US$1,700 on Canadian North, US$217 and US$144, respectively.

While regional integration is embryonic from an economic perspective, it does exist from an institutional perspective. As Richard and d'Alessandro (2018) write,

> there is certainly a close link between governance and macro-regional integration [...] From a theoretical point of view, we can indeed admit that the weaving of direct horizontal relationships between actors located in neighboring territories – across state borders, for example – contributes to the increase of interactions between these territories, to the emergence of common behaviors, norms and representations, and ultimately to the construction of integrated functional regional spaces.

The slow and uneven process of institutionalization of the Arctic region described in this chapter does not reflect the advent of a regional level of government. Institutions do, however, contribute to the densification of functional or at least discursive connections among the region's inhabitants. While the effectiveness of an institution is usually measured in terms of its ability to solve the problem for which it is set up – for the Arctic Council, for example, it was initially intended to protect the environment – another way of assessing the effectiveness of Arctic institutions would be to measure the extent to which they contribute to the construction of the Arctic as a political or even functional region. This does not necessarily imply that the territories in question must be perfectly integrated in economic, political or other terms. It means that individuals, groups and state representatives see the problems and opportunities in the region through a regional prism and that external actors also recognize the region as a whole (Escudé-Joffres 2021). The Arctic as a political project seemed to be well on its way until the outbreak of war in Ukraine, which caused a sudden halt with the suspension of the Council's activities to avoid Russia's exclusion.

The francophone literature on the Arctic region, in and of itself, is thin. While it is undeniable that the Arctic is attracting more research than it did some 15 years ago, it should be noted that this research is compartmentalized by sub-region or by theme. Regarding research by sub-regions, we can mention the work of Yvette Vaguet (2021) on the Siberian Arctic or Florian Vidal (2018) on the regional dynamics of the Barents region. Proposing a thematic approach, the geographer Éric Canobbio (2007) focuses on the entire region, but with work on indigenous populations only. Camille Escudé (2019) coordinated a book on Arctic regions and defended a thesis in 2020 on Arctic regional integration. Loïzzo and Tiano (2020) also coordinated a book on the region. Frédéric Lasserre (2019, 2020), is the leading Arctic specialist in the world of francophone geography at Laval University, whose research focuses on transportation, resources, energy, maritime issues and geopolitical issues. Laval University is also the first French-speaking pole of research on the Arctic, with, for example, Pauline Pic (2018) who works on the Arctic in a global dimension. This university hosts the ArcticNet consortium. Finally, many works of jurists focus on the Arctic region, but generally only on the maritime space in the context of theses on the law of the sea, omitting the human dimension of the territories.

Non-French research is oriented towards other topics. In the English-speaking world, it often focuses on security aspects (Huebert 2010). In the Nordic countries, it is oriented more towards the theme of governance in the region. Elsewhere, it is the roles and strategies of actors who are not members of the Arctic Council that are studied, in particular, Asian actors. We can mention the work of Mia Bennett (2004) on the extension of the region towards China. In this part of the world, the main research center is the Arctic Center at the University of Lapland, where political scientists – such as Timo Koivurova (2016) or Lassi Heininen (2019) – play a central role. Meanwhile, at Umeå University, Carina Keskitalo (2007) is interested in the role of discourses in the formation of the Arctic as a region, from a social constructivist perspective. Finally, an emerging center in Europe on the Arctic is the University of Durham, where Ingrid Medby (2018) has published a political geography dissertation on Arctic identities and their relationship to territory. Comprehensive and holistic studies are still few and far between.

12.6. References

ACIA (2004). *Impact of a Warming Arctic: Arctic Climate Change Impact Assessment*. Cambridge University Press [Online]. Available at: http://www.acia.uaf.edu/pages/overview.html.

Arctic Circle (n.d.). Arctic Circle [Online]. Available at: https://www.arcticcircle.org/about [Accessed 3 December 2021].

Arctic Corridor (2019). Growth through arctic resources [Online]. Available at: https://arcticcorridor.fi/ [Accessed 16 October 2020].

Arctic Council (1996). Ottawa declaration [Online]. Available at: https://oaarchive.arctic-council.org/handle/11374/85.

Arctic Council (2013). Arctic Council rules of procedure [Online]. Available at: https://oaarchive.arctic-council.org/handle/11374/940.

Arctic Gateway (2020). Services [Online]. Available at: http://arcticgateway.com/ [Accessed 17 November 2020].

Babin, J. and Lasserre, F. (2019). Asian states at the Arctic Council: Perceptions in Western States. *Polar Geography*, 42(3), 145–159.

Barents Euro-Arctic Region (2019). Joint Barents transport plan [Online]. Available at: https://www.barentsinfo.fi/beac/docs/JBTP2019_MAIN_REPORT_190910.pdf.

Bell, J. (2019). Greenland government rejects resumption of air link with Nunavut. *Nunatsiaq News*, 23 October [Online]. Available at: https://nunatsiaq.com/stories/article/greenland-government-rejects-resumption-of-air-link-with-nunavut/ [Accessed 16 November 2020].

Bennett, M.M. (2014). North by northeast: Toward an Asian-Arctic region. *Eurasian Geography and Economics*, 55(1), 71–93. doi: 10.1080/15387216.2014.936480.

Browning, C. (2003). The region-building approach revisited : The continued othering of Russia in discourses of region-building in the European north. *Geopolitics*, 8(1), 45–71. doi: 10.1080/714001005.

Canobbio, É. (2007). *Atlas des pôles. Régions polaires : questions sur un avenir incertain.* Autrement, Paris.

Depledge, D. (2018). *Britain and the Arctic*. Palgrave Macmillan, Basingstoke [Online]. Available at: https://www.springer.com/us/book/9783319692920.

Escudé, C. (2017). Le Conseil de l'Arctique, la force des liens faibles. *Politique étrangère, Automne*, 3, 27–36. doi: 10.3917/pe.173.0027.

Escudé, C. (2019a). *Les régions de l'Arctique*. Atlande, Paris.

Escudé, C. (2019b). Le choc des échelles : la représentation politique des territoires locaux de l'Arctique. In *Prolifération des territoires et représentations de l'Union européenne*, Brunet, S., Lebon, L., Richard, Y. (eds). PURH, Rouen and Le Havre [Online]. Available at: https://nouveautes-editeurs.bnf.fr/annonces.html?id_declaration=10000000505772&titre_livre=PROLIFERATION_DES_TERRITOIRES_ET_REPRESENTATIONS_TERRITORIALES_DE_L%27UNION_EUROPEENNE#dte-parution.

Escudé-Joffres, C. (2020a). Coopération politique et intégration régionale en arctique : naissance, développement et critique d'une région. *Belgeo. Revue belge de géographie*, 4. doi: 10.4000/belgeo.43757.

Escudé-Joffres, C. (2020b). Coopération politique et intégration régionale en Arctique (1996–2019) : construction d'une région. Thesis, Institut d'Etudes Politiques de Paris [Online]. Available at: https://hal.archives-ouvertes.fr/tel-02980869.

Escudé-Joffres, C. (2021). Coopération politique et intégration régionale en Arctique (1996–2019) : construction d'une région, naissance, développement et remise en cause d'un nouvel espace politique régional. *Carnets de géographes*, 15. doi: 10.4000/cdg.7615.

Exner-Pirot, H. and Murray, R.W. (2017). Regional order in the Arctic: Negotiated exceptionalism. *Politik*, 20(3) [Online]. Available at: https://tidsskrift.dk/politik/article/view/97153.

Flight Connections (2020). Direct flights from origin [Online]. Available at: https://www.flightconnections.com/ [Accessed 18 November 2020].

Fouche, G. and Klesty, V. (2022). On Norway's Arctic border with Russia, a town freezes ties with its eastern neighbour. *Reuters*, May 18 [Online]. Available at: https://www.reuters.com/world/europe/norways-arctic-border-with-russia-town-freezes-ties-with-its-eastern-neighbour-2022-05-11/ [Accessed 26 April 2023].

Friesen, J. (2007). Russian ship crosses "Arctic bridge" to Manitoba. *The Globe and Mail* (Toronto), 18 October [Online]. Available at: https://web.archive.org/web/20090220164851/http://www.theglobeandmail.com/servlet/story/RTGAM.20071018.wChurchill18/BNStory/National/ [Accessed 17 November 2020].

Gattolin, A. (2014). Arctique : préoccupations européennes pour un enjeu global. Information report no. 684 (2013–2014) on behalf of the Committee on European Affairs, 2 July, Sénat, Paris [Online]. Available at: https://www.senat.fr/rap/r13-684/r13-684_mono.html.

Girault, C. (2018). Institutions régionales : deux modèles principaux pour un ensemble multiforme. In *Dictionnaire de la régionalisation du monde*, Mareï, N. and Richard, Y. (eds). Atlande, Neuilly.

Gustad, R. (2001). Arctic Air flyr til Murmansk. *Nordlys*, 10 February, 17.

Hall, C.M. (2009). Changement climatique, authenticité et marketing des régions nordiques : conséquences sur le tourisme finlandais et la "plus grande marque au monde" ou "les changements climatiques finiront-ils par tuer le père Noël ?" *Téoros. Revue de recherche en tourisme*, 28(28–1), 70–79.

Hamelin, L.E. (1975). *Nordicité canadienne*. Cahiers du Québec/Hurtubise HMH, Montreal.

Heininen, L. (2019). Special features of arctic geopolitics – A potential asset for world politics. In *The GlobalArctic Handbook*, Finger, M. and Heininen, L. (eds). Springer, Cham. doi: 10.1007/978-3-319-91995-9_13.

Huebert, R.N. (2010). *The Newly Emerging Arctic Security Environment*. Canadian Defence & Foreign Affairs Institute, Calgary.

Humpert, M. (2020a). Eimskip and Royal Arctic Line commence container shipping cooperation to Greenland. *High North News*, 29 June [Online]. Available at: https://www.highnorthnews.com/en/eimskip-and-royal-arctic-line-commence-container-shipping-cooperation-greenland [Accessed 17 November 2020].

Humpert, M. (2020b). Alaska-Alberta railway clears hurdle with approval from President Trump. *High North News*, 6 October [Online]. Available at: https://www.highnorthnews.com/en/alaska-alberta-railway-clears-hurdle-approval-president-trump [Accessed 26 April 2023].

Johnstone, R.L. (2012). An Arctic strategy for Scotland. *The Arctic Yearbook 2012*, 114–129.

Juillard, E. (1962). La région : essai de définition. *Annales de géographie*, 71(387), 483–499.

Käpylä, J. and Mikkola, H. (2019). Contemporary Arctic meets world politics: Rethinking Arctic exceptionalism in the age of uncertainty. In *The GlobalArctic Handbook*, Finger, M. and Heininen, L. (eds). Springer, Cham. doi: 10.1007/978-3-319-91995-9_10.

Keskitalo, C. (2007). International region-building: Development of the Arctic as an international region. *Cooperation and Conflict*, 42(2), 187–205. doi: 10.1177/0010836707076689.

Koivurova, T. (2016). How to improve Arctic international governance. *UC Irvine Law Review*, 6(1), 83.

Landriault, M., Chater, A., Wilson Rowe, E., Lackenbauer, W. (2020). *Governing Complexity in the Arctic Region*. Routledge, London.

Lapland.fi (2020). Lapland [Online]. Available at: https://www.lapland.fi/business/tourism-industry-in-lapland/ [Accessed 24 August 2022].

Lasserre, F. (2010). Vers une autoroute maritime ? Passages arctiques et trafic maritime international. In *Passages et mers arctiques. Géopolitique d'une région en mutation*, Lasserre, F. (ed.). Presses de l'Université du Québec, Quebec.

Lasserre, F. (2018). Kirkenes, bout du monde norvégien ou place commerciale transfrontalière ? Illustration de l'effet frontière. *Regards géopolitiques – Bulletin du Conseil québécois d'Études géopolitiques*, 4(2), 19–28.

Lasserre, F. (2019). Les changements climatiques : moteur des politiques étrangères en Arctique ? *Regards géopolitiques – Bulletin du Conseil québécois d'Études géopolitiques*, 5(2), 9–17.

Lasserre, F. (2021a). Contrasting trends in regional Arctic destinational and transit shipping. In *The Arctic in World Affairs – Will Great Power Politics Threaten Arctic Sustainability? 2020 North Pacific Arctic Conference Proceedings*, Brigham, L., Corell, R., Kim, J.D., Kim, Y.H., Moe, A., Morrisson, C., VanderZwaag, D., Young, O. (eds). Korean Maritime Institute and Honolulu, East-West Center, Seoul.

Lasserre, F. (2021b). Bateaux, plateaux, arsenaux : quels enjeux géopolitiques dans un Arctique en mutation ? *VertigO – La revue électronique en sciences de l'environnement*, Special issue 33. doi: 10.4000/vertigo.29603.

Lasserre, F. and Têtu, P.-L. (2020). The geopolitics of transportation in the melting Arctic. In *A Research Agenda for Environmental Geopolitics*, O'Lear, S. (ed.). Edward Elgar, Northampton, MA.

Lasserre, F., Choquet, A., Escudé-Joffres, C. (2021). *Géopolitique des pôles : vers une appropriation des espaces polaires ?* Le Cavalier bleu, Paris.

Loïzzo, C. and Tiano, C. (2020). *L'Arctique : à l'épreuve de la mondialisation et du réchauffement climatique*. Armand Colin, Paris.

Lottero, L. (2018). L'Écosse se tourne vers le Nord. *Le grand continent* [Online]. Available at: https://legrandcontinent.eu/2018/11/26/en-attendant-le-brexit-lecosse-se-tourne-vers-le-nord/.

Mattli, W. (1999). *The Logic of Regional Integration: Europe and Beyond*. Cambridge University Press.

Medby, I. (2018). Articulating state identity: "Peopling" the Arctic state. *Political Geography*, 62, 116–125. doi: 10.1016/j.polgeo.2017.10.008.

Metz, P. and Taylor, M. (2012). Alaska-Canada railroad project. Incremental expansion project breakout. Working paper [Online]. Available at: www.interbering.com/Feasibility-Study-Reports/Alaska-Canada-Railroad-Project-by-Paul-Metz-and-Mark-Taylor-March-2012.pdf [Accessed 16 November 2020].

Nilsen, T. (2020). The dream of an Arctic railway fades as Sami herders signal "veto". *The Independent Barents Observer*, 5 March [Online]. Available at: https://thebarentsobserver.com/en/life-and-public/2020/03/arctic-railway-dream-fades-away-sami-herders-announce-veto [Accessed 23 March 2020].

Nilsen, T. (2022). Murmansk terminates sister city relations with Akureyri. *The Barents Observer*, 15 December [Online]. Available at: https://thebarentsobserver.com/en/life-and-public/2022/12/murmansk-terminates-sister-city-relations-akureyri [Accessed 26 April 2023].

Paasi, A. (2010). Regions are social constructs, but who or what "constructs" them? Agency in question. *Environment and Planning A*, 42(10), 2296–2301. doi: 10.1068/a42232.

Paquin, S. (2004a). La paradiplomatie identitaire : le Québec, la Catalogne et la Flandre en relations internationales. *Politique et Sociétés*, 23(2–3), 203–237. doi: 10.7202/010890ar.

Paquin, S. (2004b). *Paradiplomatie et relations internationales : théorie des stratégies internationales des régions face à la mondialisation*. PIE-Peter Lang, Brussels.

Parlement européen (2008). Résolution du 9 Octobre 2008 sur la gouvernance arctique (Testimony of the European Parliament) [Online]. Available at: http://www.europarl.europa.eu/sides/getDoc.do?pubRef=-//EP//TEXT+TA+P6-TA-2008-0474+0+DOC+XML+V0//EN.

Pic, P. (2018). La recomposition des enjeux stratégiques en Arctique. *Les Champs de Mars*, N° 30 + Supplément, 1, 157–165.

Polar Regions Department (2018). Beyond the Ice. UK policy towards the Arctic. Report, Foreign and Commonwealth Office, London.

Ren, Y. (2019). China in the Arctic Council. Existing problems and prospective solutions. In *Emerging Legal Orders in the Arctic. The Role of Non-Arctic Actors*, Shibata, A., Zou, L., Sellheim, N., Scopelliti, M. (eds). Routledge, London.

Richard, Y. (2018). Nouveau régionalisme : un renouvellement de la recherche sur l'intégration régionale. In *Dictionnaire de la régionalisation du monde*, Mareï, N. and Richard, Y. (eds). Atlande, Neuilly.

Richard, Y. and D'Alessandro, C. (2018). Gouvernance : mécanisme de gouvernement de la région et outil d'intégration régionale. In *Dictionnaire de la régionalisation du monde*, Mareï, N. and Richard, Y. (eds). Atlande, Neuilly.

Rottem, S.V. (2020). *The Arctic Council. Between Environmental Protection and Geopolitics*. Palgrave, London.

Salama, V. (2023). Chill settles in at a Europe-Russia border crossing. At Norway's Arctic frontier, relations with Moscow go on despite Oslo's support for Ukraine. *Wall Street Journal*, 22 April [Online]. Available at: https://www.wsj.com/articles/chill-settles-in-at-a-europe-russia-border-crossing-f5896405 [Accessed 26 April 2023].

Scottish Government (2019). Arctic connections. *Scotland's Arctic Policy Framework*, 46 [Online]. Available at: https://www.gov.scot/binaries/content/documents/govscot/publications/strategy-plan/2019/09/arctic-connections-scotlands-arctic-policy-framework/documents/arctic-connections-scotlands-arctic-policy-framework/arctic-connections-scotlands-arctic-policy-framework/govscot%3Adocument/arctic-connections-scotlands-arctic-policy-framework.pdf.

Seys, F.-O. (2019). Des régions territoriales aux régions fonctionnelles, une proposition de classification des régions de l'Union européenne. *Belgeo*, 2. doi: 10.4000/belgeo.33263.

Staalesen, A. (2019). Natural gas tankers jam on Norwegian Arctic coast. *The Independent Barents Observer*, 19 January [Online]. Available at: https://thebarentsobserver.com/en/industry-and-energy/2019/01/natural-gas-tankers-jam-norwegian-arctic-coast [Accessed 17 November 2020].

Staalesen, A. (2023). Journey's end is near for cross-border driving. *The Barents Observer*, 5 January [Online]. Available at: https://thebarentsobserver.com/en/borders/2023/01/journeys-end-near-cross-border-driving [Accessed 26 April 2023].

Stefansson, V. (1921). *The Friendly Arctic: The Story of Five Years in Polar Regions*. Macmillan, New York.

Stokke, O.S. and Hønneland, G. (eds) (2007). *International Cooperation and Arctic Governance: Regime Effectiveness and Northern Region Building*. Routledge, London.

The State Council – The People's Republic of China. (2018). China's Arctic Policy [Online]. Available at: http://english.gov.cn/archive/white_paper/2018/01/26/content_281476026660336.htm.

Vaguet, Y. (2021). Fronts et frontières en Arctique, quelle singularité ? *Géoconfluences* [Online]. Available at: https://geoconfluences.ens-lyon.fr/informations-scientifiques/dossiers-regionaux/arctique/articles-scientifiques/fronts-et-frontieres-en-arctique.

Vidal, F. (2018). Les relations entre la Norvège et la Russie en Arctique : regard et analyse sur une dynamique régionale à l'ère de l'Anthropocène. Thesis, Sorbonne Paris Cité [Online]. Available at: http://www.theses.fr/2018USPCB082.

Young, O.R. (2009). Whither the Arctic? Conflict or cooperation in the circumpolar north. *Polar Record*, 45(1), 73–82. doi: 10.1017/S0032247408007791.

13

North America: An Asymmetric Regional Integration

Christian GIRAULT

CNRS CREDA "Amériques", Condorcet Campus, Aubervilliers, France

13.1. Introduction

The northern part of the American continent, including Mexico, covers 22 million square kilometers and is populated by 500 million inhabitants. Due to its very important agricultural and mining resources and its early industrial development, this region ranks first in the world for per capita income (48,000 dollars; CEPII 2021). Its size, the meridian organization of its relief facilitating North–South exchanges, its wealth and the commonalities of a colonial history that have marked the three countries are factors favorable to regional integration. On the contrary, the conflictual affirmation of these nations in the 19th century, and the strength of nationalism in Mexico and the United States in the 20th century and the beginning of the 21st century are unfavorable factors. This chapter presents an assessment of this dialectic between nationalism and regionalism, which tends to favor the latter. Another objective of this chapter is to show that, even today, a region can be constituted despite – or thanks to – the hegemonic position of one of the countries, in this case the United States.

13.2. The region of great spaces

Apart from its geographical isolation, the region is characterized by its geophysical framework articulated in large groups: the ancient Canadian Shield and

Regionalization of the World,
coordinated by Pierre BECKOUCHE and Yann RICHARD.

the Appalachian Mountains in the east, and the high cordilleras in the west, the Sierras Madres, which prolong themselves in the Rocky Mountains up to Alaska, bordered by active fault lines on the Pacific side. In the center of the continent, the piedmonts drop down to immense plains, drained by rivers that flow into the Gulf of Mexico or the Arctic Ocean. In the northeast sector, the network of five Great Lakes represents the largest body of fresh water on the planet. In its southern part, successive isthmuses border two sea basins adjacent to the Atlantic: the Gulf of Mexico and the Caribbean Sea. The Isthmus of Tehuantepec (200 km wide) marks the threshold of Central America and the transition with the southern states of Mexico and the rest of the small Central American states up to the Isthmus of Panama. On the Caribbean Basin side, territorial continuity is replaced by old maritime and aerial circulations around the island of Cuba, the Greater Antilles and the Lesser Antilles, circulations which are still found close to the "*Tierra Firme*", with the Caribbean provinces of Colombia and Venezuela, and also the three Guianas.

It is worth emphasizing the scale of the territories involved, as well as the large extension of the exclusive economic zones in the maritime domain of three oceans. Canada has six time zones, the continental United States has five and Mexico has four. For all three countries, subjected to European colonization from the 16th century onwards, the imaginary worlds of the *Frontera*, the *Frontier*, the *Far West* or the *Great North* have remained alive. Today the challenge is more in the form of rational management of natural resources, respect for the environment and cooperation between the three states in the fields of economy, health, security and human rights.

13.2.1. *A region long isolated on the world map*

Russia and the United States, which bought the huge area of Alaska as early as 1867, face each other in the inhospitable Bering Strait area where they maintain a strategic presence to ensure the passage of circum-Arctic shipping convoys (Northeast Passage and Northwest Passage). The United States and Canada have a dispute over the latter passage, with Canada considering it to be territorial waters, while the United States pretends that several points in the passage are international straits.

The huge island of Greenland is geologically part of North America, but under Danish sovereignty. Launched in 2019, the idea of President D. Trump to buy the island seemed like one more extravagance. However, in Greenland, the United States retains the use of a well-equipped air base in Thule (Qaanaaq in Greenlandic), built during the World War in 1941 and used extensively during the Cold War. The local population, which is very small, has internal autonomy within the Kingdom of Denmark and voted for an environmentalist party that opposes rare earth mining in the south of the island in 2021.

The relative geographical isolation of North America made it safe from attack, at least by conventional means. In fact, attacks have been few and are remote in history. The major conflicts took place in the *interior* of the continent: the British against "insurgents", the United States against Mexico, the Civil War in the United States, American interventions at the time of the Mexican Revolution... For all that, when compared with other regions of the world, it is the peaceful character of inter-state relations that dominates here.

13.2.2. *A very simple political map*

As successors to the three great colonial empires – Spanish, British and French – which shared the North American space from the 16th century onwards, Canada, the United States and Mexico have adopted a transcontinental orientation from the outset (*A mari usque ad mare* according to Canada's ancient motto), with borders established along the lines of latitude.

This political architecture has advantages in terms of legibility, potential for circulation, development and territorial cooperation. The dynamics of the new countries, still evident today – which certainly do not take into account the specificities of the first peoples, the American Indians and the Inuit, decimated and then distributed in narrow reserves – were a factor of strength for the colonial and then capitalist conquest in Canada and the United States. Mexico, on the contrary, is a more complex mestizo nation, also more effaced in the contemporary period, when it suffers from serious internal difficulties (economic backwardness, marginality, crime) (Lomnitz 2021).

The presence of only three recognized official languages in the region, English, French and Spanish, encourages cooperation. Thanks to the free trade agreement between the three countries, products circulate in North America with notices in three languages, which constitutes a certain revenge for the French language courageously maintained by the French Canadian minority, long dominated by the "English".

13.3. The three partner states in North America

13.3.1. *The wars of the 19th century, the affirmation of modern nations and the delicate regional theme*

In the 19th century, young states clashed on the ground. The newly independent United States attacked the positions of the British during the Anglo-American War

(1812–1815), who retaliated with land and naval operations against the Americans (Hayes 2015). The status quo was restored by the Treaty of Ghent (1814).

In the south, the wars continued the age-old struggle between the British and Spanish for dominance of the continent. The United States supported the dissent of the state of Texas, which had declared its independence from Mexico in 1836, and incorporated it into the Union in 1845. During the Mexican–American War of 1846–1848, the Americans occupied Mexico City and dictated the Treaty of Guadalupe Hidalgo, by which Mexico renounced Texas and all its northern territories that joined the Union. These severe defeats of Mexico fueled the "anti-gringo" sentiment for many decades.

This is the opposite of a regional dynamic. Certainly, the formation of the United States itself could have been considered as a regional assembly of free states. The resolution of the Civil War (1865) wrote history differently: the Confederate States did not have the right to self-determination and were irrevocably part of the United States. The victors had imposed their national interpretation of the country.

A virulent nationalism asserted itself in the 20th century in Mexico and the United States. At the beginning of the 21st century, the presidency of the Republican D. Trump (2017–2021) participated in the return of a populist and chauvinistic movement in the United States, which could be a symptom of a nationalist polarization (Grant 2020). In Mexico, President A.M. López Obrador ("AMLO") came to power in 2018, leading a new party called the National Regeneration Movement (MORENA), which showed the strength of nationalist sentiments. The case of Canada is special because the rise of nationalism in the French Canadian community during the 20th century, mainly in Quebec, has not been counterbalanced by an English Canadian nationalism, struggling to express itself. At the beginning of the 21st century, nationalism appears to be in retreat in Canada, while identity movements of an "Indian awakening" are emerging with greater confidence.

Despite this, there has been a gradual rapprochement between the three nations since the second half of the 20th century, aided by racial desegregation, human rights legislation and growing transnational relations. There has been no confrontation on the ground since the Mexican Revolution (1910–1923), and it is unthinkable to envisage an armed conflict involving the current states.

13.3.2. *American hegemony over strategic control of the region*

The historical orientations of the three states are strongly divergent. Canada maintains the constitutional link with the British Crown, which represents both the strength of tradition and the need to distinguish itself from its American neighbor. Domestically, the federal policy of multiculturalism, implemented since P.E. Trudeau (1969), aims to overcome the division between French and English Canadians, especially since the country pursues an active policy of controlled immigration, welcoming large contingents from various regions of the world, especially countries undergoing bloody wars (Syria, Afghanistan, Ukraine). The parliamentary political system, known as the Westminster system, ensures relative fair play between the governing and opposition parties, which take turns in power.

Culturally, Mexico is tied to Latin America. President AMLO (2018–2023) cultivates this orientation, presiding in Mexico City over the Community of Latin American and Caribbean States (CELAC) Summit in September 2021 and hosting the Cuban president, an opponent of the United States. Its Constitution, inherited from the revolutionary period (1917), is original with a six-year term for the president, who cannot be re-elected. In principle, the administrative system is distributed throughout the Federation of the "Mexican United States" (the country's official name), but in reality, there is strong centralization in the country's capital, a metropolis of more than 20 million inhabitants, the most populous in North America.

As for the United States of America, it boasts the oldest Constitution in force in the world; the "fathers of the country" have constantly proclaimed the messianic mission and many politicians have taken up the antiphon of the country's "exceptionality". However, the "natural" field of American hegemony since the end of the 19th century has been the American space, with a predilection for the Caribbean Basin in the broadest sense of the word, which was once a "backyard" controlled by the Panama Canal Zone and a battery of military bases that facilitated interventions and commercial power policies, and which has now become a "security zone", considered essential by Washington.

The United States maintains strategic control over this Caribbean region from its Southern Command in Miami (*SouthCom*), through its sovereignty over a small Creole state, the Commonwealth of Puerto Rico, located in the middle of the Caribbean arc, conquered during the war against Spain (1898) and having internal autonomy, and finally through its police role, in order to control the drug trade.

For the northern sector, the North American Aerospace Defence Command (*NORAD*) based in Colorado Springs, in the geographic center of the continent, has the mission of covering the United States and Canada, militarily associated, up to the confines of the Arctic. The American bases in Alaska complete the system. Mexico is not part of this alliance, but the Mexican and US armed forces regularly participate in joint maneuvers off Acapulco and in the Gulf of Mexico.

Despite their inherent heterogeneity and historical differences, the three countries share a common federal administrative structure. In fiduciary matter, the Canadian dollar and the Mexican peso are totally linked to the US dollar, which dominates the entire American zone. There are also many commonalities in the lifestyles, legal and commercial rules, technical standards and cultural practices (media, social networks, etc.) imposed by large American companies.

13.3.3. *Neighborhood, technical and diplomatic cooperation: progress throughout the 21st century*

The three neighbors are moving closer together in a pragmatic way and are undertaking technical cooperation projects, often to the advantage of the United States, of course. The use of the waters of the Colorado River, which flows into the Gulf of California on the Mexican side, is regulated by a treaty signed in 1944 between Mexico and the United States, which have built dams on the river and developed large canals to supply a rapidly growing population in the west. This agreement is the subject of much discussion and a permanent feature of the negotiations between the two countries.

In the Northeast, the prospect of transforming the Great Lakes into an inland sea, with the possibility of deep-sea ships penetrating 3,500 kilometers into the interior of the continent, attracted attention as early as the 1930s. In 1940, President F.D. Roosevelt declared: "the United States needs the St. Lawrence Seaway for defence". The construction of the St. Lawrence Seaway, with lock systems that straddled much of the border, was also intended to promote the economic development of both countries. It was completed in 1959 (Meinig 2004).

The Bracero program, through which the United States imported contingents of seasonal Mexican labor from 1942 to 1964, was initially a necessity resulting from the World War, as men were mobilized in the armies. This program later had a great impact, as it was the source of legal and illegal migration of Mexican workers who learned about working conditions in the United States and became familiar with the language and way of life. During the Korean War, where Canada and the United States committed large numbers of troops under the banner of the United Nations,

Mexico did not participate in the operations, even though 100,000 Mexicans and Mexican-Americans were engaged on the battlefields.

However, Canada and Mexico, jealous of their identity, wish to mark their independence from their powerful neighbor and it is in international politics that it is easiest to do so. While the United States imposed a strict embargo on Cuba after the failure of the "Bahía de Cochinos" (1961), Canada and Mexico maintained diplomatic and commercial ties with the now communist island. Mexico proposed the Treaty of Tlatelolco, which prohibited the possession and use of nuclear weapons on the Latin American continent (1967); it adopted an attitude of neutrality during the Cold War and its diplomacy was close to that of the non-aligned countries.

But the weakening of the Soviet Union led to changes in the 1980s. Caught in the grip of the economic and financial crisis and agitated by domestic protest, Mexico signed the GATT agreements and was forced to move closer to the United States to solve its pressing problems. In 1994, in the midst of a financial storm (strong devaluation of the Mexican peso in December 1994) and a political storm following the assassination of political figures, it had to make an emergency appeal to the United States. As part of the rapprochement between the two economies allowed by the Free Trade Agreement (NAFTA), President W. Clinton granted a loan from the US Treasury of 50 billion dollars, pledged on oil exports.

Canada's position is also changing: now less focused on European affairs, the country is beginning to take an interest in the Caribbean and Latin America under the influence of its companies, which are eyeing the mineral resources of these countries, and also under the influence of a more active civil society that cares about the humanitarian situation in Haiti, Chile, etc. Canada joined the Organization of American States (OAS) in 1990 and then began a period of international openness with the creation of an agency of official development assistance in Ottawa, and of international research centers at various universities. Later, when the G-20 was created after the 2008 financial crisis, President B. Obama ensured that the two neighbors were part of it.

13.4. What kind of integration is taking place in North America?

In the second half of the 20th century, the idea of free trade was gaining ground among entrepreneurs and politicians. Labor unions were mostly hostile, and public opinion was undecided. Canada already had a managed trade policy with the United

States under the "Auto Pact" signed in 1964, which responded to consumer demand for products and protected the interests of the three major American manufacturers. The signing of a Free Trade Agreement between Canada and the United States in 1988 broke with decades of protectionism on both sides. Under the leadership of President C. Salinas, Mexico asked to join this movement and in 1992 signed the North American Free Trade Agreement (NAFTA) with its two partners, which took effect on January 1, 1994. The final adoption of the Agreement, whose ratification was highly contested in the United States and Canada, was only possible thanks to the conclusion of parallel agreements on working conditions and the environment. Established at that time, the Commission for Environmental Cooperation (CEC), based in Montreal, is an organization without much power for intervention, but interesting because of the studies it conducts and the reports it produces (e.g. the *North American Environmental Atlas*).

The conclusion of a trade agreement of the magnitude of NAFTA necessarily has long-term implications that go beyond the commercial sphere and affect the political, social and cultural domains. It remains a privileged axis for defining economic policies for the three countries, even though it cannot summarize their future.

13.4.1. *The consequences of NAFTA are irreversible*

All the authors of the many reviews of NAFTA have noted the deepening of trade between the three partners and the rise of cross-investment. The Canadian and US economies were already intertwined; they are becoming more so. In 2020, 63% of Mexico's foreign trade was with the United States; for Canada, the figure is 60%. The most spectacular case is the increase in relations between Canada and Mexico, two non-contiguous countries that traded very little until the 1990s. In 2020, Mexico became Canada's fourth largest trading partner, with US$25 billion in trade, and Canada became Mexico's fifth largest trading partner. The three countries are linked by production and distribution chains for energy, industrial and food products. Dense networks of oil and gas pipelines and road transport crisscross the region. A proposed merger of Canadian Pacific and Kansas City Southern is intended to strengthen the continental rail network (Myer and Fontanella-Khan 2021). After transition periods provided for in the treaty, the rules protecting some sectors such as the agricultural sector, the energy sector for Mexico or the cultural sector for Canada have been lifted. For international tourism, the three countries, which have first-class attractions, form a single market (Meinig 2014).

Beyond this observation on trade flows and investments, some authors insist on "the links that bind" in a structural relationship of dependence. The assembly plants located on the Mexican border (*maquiladoras*) existed before the adoption of NAFTA, but the Agreement reinforces the division between the ordering company in the north and the subcontractor in the south: "NAFTA locked Mexico into a strategy that ties its economic future to the United States and beyond it to the global economy" (Coatsworth 2001). In the first 15 years of the Agreement, the relocation of jobs took a massive turn due to the complete relocation of textile factories from the southern United States to Mexico. New industrial outsourcing zones are developing elsewhere in Mexico, for example, in the Bajío region (central-western Mexico). Some economists see NAFTA as a trap for the Mexican economy, which has failed to take advantage of the agreement to move upmarket and diversify its exports (Deblock 2010).

This economic subordination is compounded by the exploitation of the workforce, the majority of whom are women. The sharp difference in income between the two sides of the border has not diminished, on the contrary. The average household income was $14,000 in Mexico versus $45,300 in the United States in 2020 (OECD 2022). Average Mexican wages for unskilled labor remain 10 times lower than those paid in the United States. The wage differential is much smaller between Canada and the United States.

Finally, legal and geographic analyses show that the Agreement reflects the structural asymmetry between the three partner states and favors private investors at the expense of public institutions. In dispute resolutions, which can last for years, American bad faith and pressure allows them to win every dispute, as in the case of the long-running Canadian softwood lumber dispute (Hayter and Edenhoffer 2014).

For Mexico, the project of catching up with the income levels and participating fully in the zone of prosperity that the United States symbolizes has been a long time coming. Mexico's membership in the OECD, the club of rich countries, in 1994 did not hold out much hope of joining the "first world". This is the real paradox of integration in North America: the gap between the developed and underdeveloped world remains and is located precisely on the Mexico–US border. For Canada, the situation is different because the standard of living is close to that of the United States and the difference is essentially due to the weakness of the Canadian currency, which has been in decline for several decades. The majority of Canadians do not look to the United States as a model because they wish to maintain their civic values and, above all, avoid the influence of crime, which is so widespread in the United States.

13.4.2. *An agreement badly affected by September 11 and cartels, but eventually solid*

The events of September 2001 dealt a severe blow to open borders. The United States refused to accept any help from its Mexican and Canadian neighbors who had spontaneously offered their services. Nor did Washington call on the Reciprocal Assistance in American Space Treaty (TIAR treaty of 1947), which showed a certain disdain for its partners. The lines of trucks grew longer at the border crossings. The American territory was closing in on itself. The borders began to "thicken" due to the construction of several physical lines of defense: barbed wire, fences, trenches, radar, electronic detectors. Mexican residents can no longer return to their country of origin. In 2003, the United States created a genuine Ministry of the Interior (significantly named the Department of Homeland Security), which brought together numerous services such as the forces fighting terrorism, customs and immigration services. This security orientation is based on the misconception that there are terrorist threats on the borders and suspicion of visitors and immigrants from neighboring countries. There is no question of cooperation between states at this level.

Concerned about the weakness of Mexican forces overwhelmed in the fight against trafficking, the United States provided financial assistance starting in 2007 for the training of Mexican police and military (a program called the Mérida Initiative). President F. Calderón had indeed declared war on drug trafficking, but the United States had little confidence in the Mexican military (Rouquié 2013). Fifteen years later, the toll of this almost lost war is appalling as the trafficking networks, called cartels in Mexico, constrained in their actions by increasingly sophisticated detection measures, have become more powerful and more violent, and have expanded their criminal actions. The issue of drug and human trafficking on the borders of the United States has become the greatest challenge to the security of the continent. Under the mandate of President AMLO, the Mexicans are pretending to reject US aid as humiliating interference. A new fund called the "Bicentennial" is to replace the Mérida Initiative (October 2021). In July 2022, the US and Mexican presidents agreed on a plan to improve border infrastructure and increase investment in Mexico.

13.4.3. *The renegotiation of NAFTA under Donald Trump*

President D. Trump blew hot and cold on the agreement. From the beginning of his term, he said that NAFTA is "the worst trade agreement in existence". In addition to hurling insults at Mexicans, he criticized the treaty, based on a simplistic economic analysis, the large trade deficit that the United States has with Mexico and

Canada. He announced the creation of a "wall" along the Mexican border to contain illegal immigration. This wall already existed in part, but the work was completed with physical and electronic defenses along a long stretch.

Then came the beginning of a trade war in 2018 with the imposition of tariffs on Canadian, Mexican and European steel (with a 25% tax) and on aluminum (10%) to protect American products and workers. In retrospect, these measures can be interpreted as pressure to renegotiate the Agreement, pushed by the United States. Canada was virtually put on notice to join the Agreement already reached with Mexico. Signed in 2018 and entering into force in 2020, the new Canada–US–Mexico Agreement (CUSMA) in fact maintains the essence of the previous treaty and reinforces the asymmetry between the three partners by tightening the rules of origin in order to relocate production, particularly of automobiles, to the United States (Ciuriak et al. 2020).

Experts' predictions of a slight decline in the GDP of the three countries depending on the implementation of the Agreement have been overtaken by the consequences of the Covid-19 health crisis, which has plunged the economies of the three countries. The blows of the Trump administration have both strengthened the Agreement and consolidated the asymmetry between the partners. As the borders closed again, the vulnerability of the two minor partners was apparent in all its extent. Only the United States could offer vaccines manufactured in part on its soil; the vaccine distribution policies pursued by the Trump and Biden administrations in 2020 and 2021 were timid. In terms of the number of contaminations and deaths, the balance sheet appeared very unfavorable for the region, compared to the European Union or China.

13.4.4. *The reality of population integration and ideological battles over migration*

Trade treaties do not address the issue of migration. Researchers denounced very early on the basic contradiction of selective globalization. Goods, well identified in the nomenclatures, free of rights, cross-borders easily. But people have to obtain expensive visas, and are then controlled, chased and sometimes harshly turned away (Castro 1999). This issue is fundamental because it is at the heart of the three countries. Canada and the United States are traditional immigration countries. Their policies have varied historically; the flows of immigrants to them came initially from Europe and then diversified with significant flows from Asia and Latin America. Until recently, Mexico was essentially a country of emigration. It did receive a contingent of Spanish political refugees after the defeat of the Republic, but this was an exception.

In recent times, migration policies in all three countries have been evolving towards tighter restrictions, careful selection of migrants, and sometimes repression and violation of rights. Many factors come into play in the definition of migration policies and their implementation. While the gradual aging of local populations should encourage the immigration of managers and workers to fill unfilled jobs, there is still a reluctance to open border crossings. The tightening of US entry requirements began in the 1980s under the two terms of G.W. Bush and continued under the two terms of B. Obama (Cohen 2012). The events of 2001, the economic and financial crisis of 2008–2009, D. Trump's openly xenophobic attitude, and the effects of the health crisis reinforced these restrictive policies.

The consequences of these policies are most apparent in border areas. Evidence and studies show that in the past century, border regions lived in a certain harmony that was expressed in numerous cross-border exchanges. For example, in the Pacific region, geographers at Simon Fraser University and Eastern Oregon State College point to the good understanding that existed between the province of British Columbia and the states of Oregon and Washington, symbolized by a Peace Arch erected on the border in 1921 (Evenden and Turbeville 1992). At the turn of the century, observers even envisioned a hypothetical obliteration of the border in the west in a new "*Amexica*" entity (*Time Magazine* 2001). Similarly, Professor C. Fortuné, who has resided in Douglas, Arizona on several occasions, attests to the nostalgia of the border town's residents who miss the ease with which we could cross the line and form friendships "on the other side" (Fortuné 2020). This conviviality was in fact broken and C. Fortuné even speaks of "border insecurity", insofar as the approaches to the border, especially on the Mexican side, are increasingly dangerous because they are controlled by drug cartels.

Since 2008, Mexico has been crossed by "caravans" of Central American migrants, followed by large groups of Cuban, Haitian and African migrants heading for the US border. This has increased pressure on the Mexican government to drive out the migrants and control its southern border. This effectively shifts the border line from the Rio Grande to Tapachula and the Rio Suchiate on the Chiapas–Guatemala border to the south (Faret 2018). Mexico, whose nationals have been discriminated against in the United States for many decades, often behaves brutally towards migrants itself (Calderón Chelius 2009). In the United States, a nation made up largely of migrants where the *melting pot* of these migratory currents was supposed to take place, racist practices from long ago are being renewed (Grant 2020).

A synthetic assessment of the migratory movement in the region shows the real possibilities of integration of migrants over two generations, certainly in Canada, but less easily in the United States. The remittances sent by Mexican workers represent sums of around 40 billion dollars per year, which support millions of poor families in Mexico. In the United States, however, there are risks of restriction – not to say segregation – based on a subtle but real ethnic segmentation. Indeed, the minority of "Mexican-Americans", categorized by authorities as "Hispanics", many of whom (about 10 million) are in the country illegally, constitute a vast underclass of the US population (Simon 2015).

13.5. Conclusion: the dilemmas of hyperpower in international competition

In the vast spaces of the North American region, "*continentalization*" is tending to strengthen, while the hegemony of the United States is becoming more influential, as shown by the recent episode of the renegotiation of the Regional Trade Agreement. The asymmetrical relationship between the partner states is even more evident today than it was in the 20th century. The effects of the September 2001 attacks and the Covid-19 health crisis have accentuated the imbalances. The case of North America shows that regional integration can progress despite, or because of, the hegemonic position of a powerful country in relation to its neighbors.

There is great cohesion among the economies, made up of intense trade and great logistical and strategic interdependence, but the creation of a North American bloc is unlikely because the three countries are subject to contradictory, violent, sometimes erratic internal political dynamics. D. Trump's tenure (2017–2021) has shaken confidence inside and outside the United States, and President J. Biden does not appear to have all the political wherewithal to restore American leadership. In addition, President AMLO's nationalist rhetoric in Mexico makes coordination with Washington and Ottawa difficult. In contrast, Canada, with its traditional two-party system and political alternations in the middle of the political spectrum, appears to be a pole of stability and tranquility (with a very low crime rate compared to its neighbors). It is difficult to see Canada and Mexico giving up their sovereignty to allow the United States to lead a possible bloc. North American regional integration, including Mexico, has become a reality, but the advent of a "North American community", once envisaged by CFR experts, is no longer on the agenda (Manley et al. 2005).

Finally, in the competition between regions of the world, the proximity and privileged links with the Latin American–Caribbean region have long served US and Canadian firms in the areas of trade and investment. But this region in the south, which is very heterogeneous, poorly integrated and subject to contradictory and sometimes destructive forces (authoritarian regimes, social marginalization, violence and criminality), is now a point of weakness, due to the almost general recessionary effects of Venezuela's economic and social collapse, for example, which are causing intense migratory pressures on the borders. Thus, we should not rule out the hypothesis that North American integration will continue to strengthen faster than problematic Latin American integration.

13.6. References

Calderón Chelius, L. (2009). Conférence d'inauguration de la deuxième semaine nationale de la migration au Mexique. *Problèmes d'Amérique Latine*, 75, 93–107.

Castro, M.J. (ed.) (1999). *Free Markets, Open Societies, Closed Borders? Trends in International Migration and Immigration Policy in the Americas*. North South Center Press, University of Miami.

CEPII (2021). *L'économie mondiale 2022*. La Découverte, Paris.

Ciuriak, D., Dadkhah, A., Xiao, J. (2020). Quantifying CUSMA: The economic consequences of the New North American Trade Regime. Working paper, C.D. Howe Institute, Toronto.

Coatsworth, J.H. (2001). Mexico and the United States *ReVista. Harvard Review of Latin America*, Fall, 3–5.

Cohen, J. (2012). *À la poursuite des illégaux. Politiques et mouvements anti-immigrés aux États-Unis*. Éditions du Croquant, Bellecombe-en-Bauges.

Commission de Coopération Environnementale (n.d.). Atlas environnemental de l'Amérique du Nord. *Commission de coopération environnementale* [Online]. Available at: http://www.cec.org/fr/atlas-environnemental-de-lamerique-du-nord.

Deblock, C. (2010). L'Alena, tremplin ou trappe ? Research paper, Centre d'études sur l'intégration et la mondialisation, Institut d'études internationales de Montréal, Université du Québec à Montréal.

Evenden, L.J. and Turbeville III, D.E. (1992). The Pacific coast borderland and frontier. In *Geographical Snapshots of North America*, Janelle, D.J. (ed.). Guilford Press, New York.

Faret, L. (2018). Enjeux migratoires et nouvelle géopolitique à l'interface Amérique Latine – États-Unis. *Hérodote*, 171, 89–105.

Fortuné, C. (2020). Sécurité frontalière, insécurité locale dans les borderlands États-Unis/Mexique. Étude de Douglas (Arizona) et Agua Prieta (Sonora). PhD Thesis, Université Sorbonne Nouvelle, Paris.

Grant, S.M. (2020). Toward a topography of national trauma: Mapping the past onto the future of American nationalism. *Nations and Nationalism*, 26(2), 366–387.

Hayes, D. (2015). *Historical Atlas of Canada*, Revised edition. Douglas and McIntyre, Madeira Park.

Hayter, R. and Edenhoffer, K. (2014). Trade disputes, dispute settlement mechanisms, and local developments. *The Professional Geographer*, 66(4), 631–640.

Lomnitz, C. (2021). México: El tejido roto. *Nexos*, 1 April [Online]. Available at: https://www.nexos.com.mx/?p=54499.

Manley, J.P., Aspe, P., Weld, W.F., d'Aquino, T.P., Rozental, A., Pastor, R.A., Lawson, C.H. (eds) (2005). *Building a North American Community*. Council on Foreign Relations, New York.

Meinig, D.W. (2004). *The Shaping of America. A Geographical Perspective on 500 Years of History. Volume 4: Global America 1915-2000.* Yale University Press, New Haven.

Meyer, G. and Fontanella-Khan, J. (2021). Rail deal builds on Canada, U.S. and Mexico ties. *Financial Times*, 22 March [Online]. Available at: https://www.ft.com/content/13d92e28-eac1-4139-b36f-823c40e6cd44.

OECD (2022). Better Life Index [Online]. Available at: https://oecdbetterlifeindex.org/fr.

Rouquié, A. (2013). *Le Mexique, un État nord-américain*. Fayard, Paris.

Simon, P. (2015). Review of Richard Alba and Nancy Foner, strangers no more: Immigration and the challenges of integration in North America and Western Europe. *Population*, 70(3), 636–639.

Time Magazine (2001). Vanishing Border. Welcome to Amexica. Special Issue, 11 June.

14

Latin American Integrations

Sébastien VELUT

EHESS/Géographie-cités, Institut Convergence Migrations, Paris, France

14.1. Introduction

The name Latin America appeared in the mid-19th century, following the independence of the Spanish colonies, along with the idea of political integration of the nations resulting from the dismemberment of the Empire. Two centuries later, the term is widely used. It was taken up by the United Nations, whose regional divisions consider a large Latin American and Caribbean region and which has a specialized agency with the Economic Commission for Latin America (CEPAL) created in 1948. This was renamed in 1984 as the Economic Commission for Latin America and the Caribbean, a precision that is not without importance, and which shows that the usual name is considered imprecise or unsatisfactory.

Many institutions, scientific journals and cultural events have adopted it. It now refers to all the states that emerged from the Spanish and Portuguese empires. In geography, the term has become established as evidenced by the publication in the *Géographie Universelle* directed by Roger Brunet of a volume entitled Latin America (Bataillon et al. 1994), whereas the previous universal geographies used other divisions. This was the case, for example, in the *Géographie Universelle* directed by Paul Vidal de La Blache and Lucien Gallois, which distinguished Central America and Mexico (Sorre 1928) from South America (Denis 1927).

Is this convergence sufficient to carry an integration project? Despite the efforts made and the countless speeches on integration, Latin America was no more integrated in 2020 than it was in 1820, whatever the meaning of the term integration.

Regionalization of the World,
coordinated by Pierre BECKOUCHE and Yann RICHARD.

In the early 2020s, regional integration is at its lowest level in half a century. Ideological rivalries between states and leaders are adding to the deep crisis of the integration institutions set up in recent decades, such as MERCOSUR and UNASUR (Union of Southern Nations), despite the more recent advances of CELAC (Community of Latin American and Caribbean States). The economic projects that justified their creation have not brought the hoped-for growth, which is sustained much more by demand from distant partners than by trade within the Region[1]. In 2020, the Covid-19 pandemic reinforced nationalist reflexes, justified the closure of borders and highlighted the powerlessness of continental institutions to deal with a common problem.

We can assume that this crisis of integration is not only cyclical, but that it shows that the organization of the Latin American region can be thought of as being between several models: on the one hand, advanced integration, at all levels, and on the other, the strength of nations that nevertheless need instances of dialogue and minimal coordination. Integration in the sense of unification is not an inevitable historical destiny, even though the states share a certain number of environmental, economic and security issues. It is this tension between fragmentation and integration that we explore here, in order to understand the strength of an idea that, in the end, has never been achieved.

Firstly, we present the genesis of the idea of Latin America, which began with the independence of the Spanish colonies, in favor of a political integration of the continent. This moment is essential because it is a founding one. Despite the importance of this integration thinking, Latin America remains divided and the integration institutions that have multiplied since the 1960s have not managed to overcome these divisions. Nonetheless, non-state actors contribute to the existence of a certain type of integration in a large fragmented entity, which shows that the integration of this large space is still relevant.

14.2. Naming a large region to integrate it

The term Latin America was coined in the 19th century and carries a geopolitical charge from the outset: to designate a part of America that is not English-speaking. However, this imprecise adjective carries several contradictory interpretations in its origins as well as in its meanings. The expression was invented in France by South American exiles and found an excellent echo chamber during the Second Empire,

1. We use the term "Region" with a capital "R" to emphasize that it is a large regional entity composed of several neighboring states. The name "Region" avoids confusion with a classic intra-state region.

when Napoleon III sought to base his foreign policy on a privileged relationship with the Latin peoples, with the Emperor's interventions in Italian unity and his Spanish marriage. The expedition to Mexico in 1861–1863 to place his nephew Maximilian of Austria on the throne was part of this logic, imposing the notion of Latin America from the European powers wishing to make a place for themselves there.

However, a few years earlier, the first uses of the term "Latin America" referred to the need to unify the independent republics of South America in the face of the growing power of the United States. This is the case for the Colombian poet José Maria Caicedo in an 1857 poem entitled *Las Dos Américas* – The Two Americas, which contrasts English-language America, i.e. the United States, with Latin America. It is also used, in a similar sense, by the Chilean journalist and writer Francisco Bilbao in a speech given in Paris in 1856 on the occasion of an unofficial congress of South American nations.

Under their respective pens, the expression replaced that of "Latin races" found in other contemporary texts, to call on the states of the subcontinent to unite against the expansionism of the United States. However, in the middle of the 19th century, the latter extended its hold westward and southward to the detriment of Mexico, which lost a large part of its territory, including California, in 1848 with the Treaty of Guadalupe Hidalgo. The United States also supported the adventurer William Walker, who took power in Nicaragua between 1855 and 1857, before taking an interest in Panama as a strategic point of passage between the two oceans, and causing its division with Colombia in 1903.

In the states that had gained independence from the Spanish Empire a few decades earlier, the expansionism of the United States was perceived as a serious threat, and calls for unity were made to form a collective front. The expression Latin America thus gives new meaning to the project expressed by the *libertador* Simon Bolivar in his Letter from Jamaica (1815), which, after reviewing the political situation of the independent states, warns against "the grandiose idea of forming a single nation out of the whole new world" because too many specific features separate the states, but calls for the formation of a union:

> how beautiful it would be if the Isthmus of Panama were for us what Corinth was for the Greeks! [May we] one day install there an august congress of representatives of republics, kingdoms and empires to treat and discuss the high subjects of peace and war

This vision is still often evoked to justify integration projects, even though Bolivar does not speak of Latin America, but of South America, and he does not take into account Brazil, which has not yet cut its ties with the Portuguese metropolis.

The Bolivarian heritage is periodically invoked and reactivated, as were the myths in ancient Greece (Veyne 1983). The Bolivarian Alliance (ALBA) was launched under the impetus of Hugo Chavez, President of Venezuela from 1999 to 2013, with the aim of uniting Latin America. More broadly, various names are used to reinforce the idea of unity in the face of external threats. The Bolivarian heritage is then claimed to recall this common past. This is the case of the formula *Nuestra América*, used by the Cuban poet José Marti in 1891, which sets aside the idea of Latinity, imposed from outside, and promotes his own version of a united South America facing the United States and seizing its historical destiny.

Although the notion of Latin America has become widespread, its definition remains elusive, and other terms compete with it. In the United States, where many universities have centers for Latin American studies, other terms are used in international politics, such as the Western Hemisphere, which refers to the entire American continent (often referred to in the plural as the Americas). In political terms, the Bureau of Western Hemisphere Affairs of the Department of State oversees all diplomatic relations with Latin America. In Spain, the term "Ibero-America" is more commonly used, as it directly recalls the past of the Spanish Empire, while the term "Latin America" seems too *French*. The Ibero-American Secretariat (SEGIB) is an international organization, based in Madrid, which includes all the former territories of the Empire, as well as Brazil and Portugal. Therefore, there is no convergence, neither on the name, nor on the limits, nor on the political project. Depending on whether one is referring to a large sub-continental entity with territorial continuity south of the Rio Grande, to the historical legacy of the Spanish Empire or to a geopolitical situation or posture vis-à-vis the great powers, particularly the United States, the meaning given to integration varies.

14.3. National and regional integrations

Latin America has a rich political and intellectual tradition that is part of what is commonly called "Latin American thought", i.e. a collective effort to think about the singularity of this great region, which is crossed by problems common to the different states. Its origins go back to independence and the reflections on the most appropriate political regime. Simon Bolivar, in a speech delivered in Angostura (today Ciudad Bolivar in Venezuela) in 1819, embodies this reflection, which was considerably enriched afterwards. This Latin American thought continues up to the current proposals around decoloniality and feminism, through the very rich reflection on the conditions of development and the relations between state and society in a context marked by the persistence of very great inequalities. This moment of reflection, which took place in the second half of the 20th century, is particularly rich and its contributions to the understanding of the problems of development go far beyond Latin America. Although shared by many observers, this intellectual tradition is ambivalent when it comes to integration: on the one hand, by highlighting common issues, it justifies a rapprochement in order to better confront them; on the other hand, the focus on the question of the state, for example by authors such as the Bolivian Raul Zavaleta, tends to value the national scale in its specificities to the detriment of a supranational approach.

The history of the independent states of Latin America in the 19th and 20th centuries has been marked by calls for unity as well as by a series of confrontations to define borders and nations. Border disputes and wars began with independence and continue to this day, although armed confrontations occurred mainly in the 19th century, with Latin American states subsequently striving to find negotiated solutions. These conflicts are an integral part of the process of national construction through the affirmation of states and their armies, of progressively defined and delimited territories and of nations whose identities are to be invented.

Several important conflicts occurred in the second half of the century: the War of the Triple Alliance (1865–1870), which saw Paraguay crushed by the coalition of Argentina, Brazil and Uruguay; and the War of the Pacific, which opposed Chile to Bolivia and Peru (1879–1884). The victory of Chile against Bolivia and Peru allowed the expansion of Chilean territory towards the Atacama Desert, depriving Bolivia of its access to the sea. This subject is always sensitive for the Bolivian state which claims by various means the return of the irredent provinces or, at least, the sovereignty on a part of the Pacific coastline (Perrier Bruslé 2013).

The war of the Triple Alliance led to the annexation of part of the Paraguayan territory to the benefit of Brazil and Argentina, but above all to the decimation of the Paraguayan male population, mobilized in an unequal struggle, and which was again engaged in the Chaco war, this time against Bolivia in the 1930s. For its part, Brazil expanded its territory from the end of the Empire to the beginning of the Republic, and established the idea of Brazil as an expansionist power (Droulers 2001). These border redefinitions are not exceptional. Europe has experienced them, much more deadly, but Latin American nationalisms continue to feed on them, making obstacles to rapprochement. For example, Chile and Argentina are in severe competition to control Patagonia and define its borders, and are on the verge of conflict on several occasions, despite the existence of a peace and friendship treaty signed in 1881 to establish the borders. In 1982, the last episode of tension gave rise to the arbitration of Pope John Paul II. This intervention, as well as the appeals to the International Court of Justice in The Hague, shows the efforts of the Latin American states to avoid open conflicts and preserve the peace of the continent, in a context of inter-state rivalry.

With the drawing of borders and the loss of territory, the ideological construction of national identities made through differentiation with the neighbors, to whom were lent the traits of a supposed national character in order to better assert by contrast the national identities under construction. National historical narratives have more readily portrayed border countries as enemies than as allies. It is only at the margins that the institutions of integration have tried to promote processes of exchange and rapprochement to overcome these differences.

Even though the constitution of national stereotypes is primarily a matter of discourse, while identities are fluid and populations mobile, the fact remains that Latin American populations are diverse and think of themselves on national scales consolidated for two centuries by educational systems and other institutions of socialization. An inventory of differences would be inexhaustible. The major opposition of official languages is not reduced to the Spanish–Portuguese couple, since some Central American and Caribbean states are partially or totally English-speaking and Amerindian languages are also widely practiced and recognized – this is the case of Guarani in Paraguay.

In addition, there is the diversity of settlement patterns, the origins of migrant populations and the characteristics of Amerindian populations, all of which have been united, at least in part, by a national integration project. In the case of Argentina, for example, national integration was conceived around a political project of occupation of the national territory for its valorization by migrant populations, mainly from Italy and Spain. In Brazil, the variety of migratory origins is even greater, with the historical presence of Amerindian populations, Afro-descendant populations deported as slaves, as well as the arrival of migrants of various origins, including from Japan. Brazil has had to invent a national narrative highlighting this diversity of origins, around a mythical narrative of miscegenation that in fact masks considerable inequalities (Droulers 2001). Both Peru and Mexico recognize the existence of more than 50 indigenous peoples in their territories, who must also be integrated into the nation, and make particular reference to their rich pre-colonial past.

The diversity of populations and the expansion of territories lead Latin American states to be much more concerned with internal integration than with regional integration. In the 20th century, nationalist projects and states were strengthened. The institutions of socialization, starting with the school, were systematized, while the occupation and planning of national territories were affirmed. The planning was designed to better connect the remote regions to the capital and to open up new territories for exploitation: inland regions of Brazil won by pioneer fronts (Droulers 2001; Théry 2012), Argentinean and Chilean Patagonia (Grenier 1988), lowlands in Colombia, southern Mexico (Revel-Mouroz 1972). The most telling case is that of Brazil, which from the 1950s onwards built roads into the Amazon, above all to assert its sovereignty thereby encouraging migration. It was planned to build a road north of the Amazon, which was never built, forming a strategic ring road to defend the national territory against external aggression – which did not happen. Thus, over the long period of the 19th and 20th centuries, the states sought more to strengthen their coherence and their borders, to integrate the national territory and the people who lived there, rather than to integrate into a great Latin American whole.

As a result of these efforts, contemporary integration remains limited by distance and insufficient infrastructure. Latin America covers about 20 million square kilometers, compared to about 4 million in the European Union, with a population of 600 million, compared to 400 million, 80% of whom are urbanized. It is therefore a very large area, with a low average population density, but with urban and metropolitan settlements, distant from each other, generally in coastal areas. Land interconnection has always been a problem, even within national borders, between coastal urban centers and inland regions and even more so between national spaces separated by low-density regions and relief. This is the paradigmatic case of the Amazon, shared between six states, but difficult to cross in the heart of South America. The member states of the Amazon Cooperation Treaty organization address together the environmental issues of the Amazon, but have not made it a space of integration. Similarly, the Andes, whose crossing requires very high altitude passages, despite multiple tunnel projects that have not been completed, or, in the American isthmus, the Darien forest region between Panama and Colombia remain major obstacles.

In the early 2000s, the Regional Initiative for South American Integration (IIRSA) was launched in Brasilia during a meeting of heads of state. The aim was to build a number of integration corridors on a continental scale (Velut 2009), by combining roadways and waterways for the transport of people and goods. The Puebla-Panama initiative, supported mainly by Mexico and later renamed the Mesoamerica project, aims to connect Mexico with the southern part of the American Isthmus (Medina 2019). Twenty years later, considerable work has been carried out in South America, with the strengthening of several of these axes, the improvement of road axes and border crossings. This is the case of the Capricorn axis, which follows the Tropic from the metropolis of São Paolo and the port of Santos in Brazil to Antofagasta in Chile, passing through northern Argentina and the Andes Mountains. This is also the case of the São Paolo–Buenos

Aires–Santiago–Valparaiso axis, which links the three main metropolises, although the major investments intended to facilitate the crossing of the Andes in all seasons have not been made. But these achievements are far from covering the whole plan formulated at the beginning of the century. Supported mainly by Brazil and financed by its National Bank of Economic and Social Development (BNDES), this project appeared as the instrument of Brazilian influence in South America, to make Brazil a sub-regional hub (Scholvin and Malamud 2020). But IIRSA also appears to be a vehicle for elite corruption, as highlighted by the so-called "*lava jato*" ("express wash") investigation. This investigation revealed that Brazil's large public works groups systematically bribed political decision-makers in neighboring countries to advance their projects. The initiative of integration through infrastructure is thus associated with serious legal and political problems and projects are stopped. Thus, even in a geopolitical context that was particularly favorable to integration, from the late 1990s to the mid-2010s, a period in which major projects emerged, achievements remained behind and integration incomplete.

14.4. Advances and impasses of contemporary political projects

From the second half of the 20th century onwards, several supranational organizations were set up to promote Latin American integration, but generally on small perimeters compared to the whole of Latin America (Dabène 2014). Integration institutions have multiplied, with different perimeters and objectives, giving rise to a complex institutional tangle. They are added to the Organization of American States (OAS), which brings together all the countries of the continent, from Canada to Tierra del Fuego. The United States plays a dominant role in this organization, which is why the other states of the continent are suspicious of it. By creating smaller integration bodies, the Latin American states are seeking to have spaces for interaction with their neighbors and partners, without the presence of the United States, mainly for economic development purposes, as well as with political objectives. These include having forums for dialogue between member countries and being able to conduct joint negotiations with the rest of the world.

On the economic front, the European model has inspired several initiatives. The oldest is the Central American Common Market, founded in 1960, which groups together the countries of the isthmus, with the exception of Panama, which is still very dependent on the United States, and Belize, which remained a British colony until 1981. The Andean Community of Nations (CAN) was created in 1969 and extended to all Andean countries, including Chile, which withdrew in 1976, and Venezuela, which left in 2006. MERCOSUR, the Common Market of the South, was launched in the 1990s by Brazil, Argentina, Uruguay and Paraguay, with the aim of creating a true common market between member countries. It quickly attracted many associate members. Venezuela applied to join in 2012, at a time when MERCOSUR's economic objectives had already been revised downwards after the economic crisis that hit Argentina in 2001–2002.

These economic integration initiatives were designed with a view to industrialization and economic complementarity in order to stimulate trade of manufactured goods between member countries. They have allowed the consolidation of certain sectors, such as the MERCOSUR automobile industry between Argentina and Brazil, and facilitated trade of foodstuffs. But they have not had the expected effects on industrial development as a whole and the structuring of territories (Guibert et al. 2009). Latin American states trade much more with the rest of the world than with their neighbors, especially in the 21st century, marked by the growth of raw material exports and the import of manufactured goods with a major partner: China. The similarity of the Latin American economies limits internal trade and the usefulness of a mechanism such as a common market. The states have been very cautious in liberalizing trade within the region and some have preferred to negotiate free trade treaties with extra-regional partners. Chile has signed more than 20 free trade treaties, notably with China, the United States and South Korea, as well as with the main Latin American economies. Peru, Colombia and Costa Rica are following the same trajectory, seeking to multiply international trade agreements. Brazil and Argentina are not following the same strategy of agreements, but their main trading partners are China and the European Union. As Girault shows in this volume, only Mexico, with the NAFTA agreement, is in a different dynamic.

The opening up of economies to globalization has led to the search for new multilateral cooperation agreements with the aim of developing trade with Asia, with the creation of the Pacific Alliance, which brings together Chile, Peru, Colombia and Mexico to promote trade relations with the countries of the Pacific, without aiming for the very ambitious objectives of a common market. Chile, Mexico and Peru also joined the Trans-Pacific Partnership project launched under the Obama administration, and the subsequent CPTPP agreement, in which China plays the leading role. Colombia has also applied for membership. These countries claim to be "open regionalism" (Diamint 2013), in other words, multiple and nonexclusive alliances, combining relations with neighboring countries and economic openness to globalization. Economic partnerships are privileged and regional or sub-regional integration is no longer a priority objective, and the possibility of creating new norms through integration seems remote, which leads to talk of "liquid regionalism" (Mariano et al. 2021).

Regional integration remains a political issue. Some trade agreements have been extended by political mechanisms. The Andean Community of Nations and MERCOSUR, for example, have regional parliaments (the Andean Parliament and Parlasur), which have no real decision-making power, but which constitute forums for dialogue. In the early 2000s, political integration was revived with the project, proposed by Venezuela, of a Bolivarian Alliance of Latin America (ALBA) to respond to the free trade area of the Americas project wanted by the United States and refused by most of them. The Bolivarian alliance brought together only a small number of countries, all of which were very committed to opposing the United States: Venezuela, Cuba, Bolivia, Nicaragua and several Caribbean states. Ecuador was part of it and then withdrew.

The creation in 2008 of UNASUR (Union of Southern Nations), bringing together for the first time all the states of South America, mainly at the initiative of Brazil, allowed the latter to consolidate its continental leadership by accepting to share it. UNASUR had a promising start, with the support of all countries. It was led by important political figures, such as former Argentine president Nestor Kirchner, who became secretary general a few months before his untimely death in 2010, and then former Colombian president Ernesto Samper, secretary general from 2014 to 2017. UNASUR gained political weight in areas of common interest such as the defense of democracy and dialogue among member countries, and was entrusted with projects of collective interest such as integration through infrastructure and energy.

But this strength has become a weakness. Consistent with the political projects of the left-wing governments of the early 21st century, UNASUR has not been supported by the right-wing and center-right governments that succeed them. Seven countries, including Colombia, Chile, Argentina and Brazil, withdrew from UNASUR in 2018, which lost its raison d'être. This organization had for the first time a perimeter covering all of South America, with the exception of French Guiana, a coherent geographical entity to think about the development of a large space.

At the beginning of the 21st century, heads of state with progressive values came to power. Some have a personal history of opposition to military dictatorships. The ideological closeness of presidents such as Michelle Bachelet (Chile), Luis Inacio da Silva (Brazil), Rafael Correa (Ecuador), Nestor and Cristina Kirchner (Argentina), Hugo Chavez (Venezuela), Evo Morales (Bolivia), José Mujica (Uruguay) or Fidel and then Raul Castro (Cuba) should not be overestimated. Nevertheless, they all share a common desire to strengthen political integration between states, without abandoning national sovereignty, in order to carry weight in international forums and in the face of the United States. After the disappearance of UNASUR, a new forum for dialogue, the Community of Latin American and Caribbean States (CELAC), has emerged, without having the same ambitions in terms of integration. Its originality lies in the inclusion of all Latin American states and its extension to the Caribbean, with the participation of several English-speaking island states. CELAC positions itself as a global interlocutor, particularly in relations with China and the European Union. However, at the last meeting of CELAC, held in Mexico City in September 2021, Mexican President Andrés Manuel Lopez Obrador proposed to relaunch continental integration in order to create a mechanism capable of replacing the Organization of American States, which is too dependent on the United States.

This succession of integration initiatives, with varying justifications, objectives, perimeters and institutions, makes Latin America a laboratory for regional integration, even though the experiments carried out in this laboratory do not produce the expected results and raise questions about the very meaning of integration.

14.5. True integration by the actors

The idea of Latin American integration has been around for more than two centuries, but this expression can be understood in different ways. The actors of society, including companies, populations and social organizations, project themselves in a shared Latin American space that they contribute towards building.

Despite the marked differences between countries, Latin America constitutes a large entity that is partially unified by economic actors and populations. Economically, international companies known as multi-Latinas operate on a continental scale. In the retail sector, this is the case of the Chilean group Cencosud, which has operations in Peru, Colombia, Argentina and Brazil, where it has acquired the Pão de Açucar supermarket network. In the air transport sector, with the LATAM group, resulting from the merger between the LanChile company and the Brazilian company TAM. The Brazilian public works company, Odebrecht, had expanded its activities throughout the continent, before sinking in the scandal triggered by the investigation into the so-called "lava jato" corruption. More generally, multinational companies operate on a Latin American scale and deploy their productive networks there.

The geography of transport and energy networks reflects these logics. In the 1990s, South America developed gas and electricity exchange infrastructures and improved international road routes. In the case of energy, there is a logic of optimizing the continent's resources to serve consuming regions. However, in the absence of sufficiently restrictive legal and commercial frameworks, trade has been reduced, leading states to seek national solutions for their energy supply (Carrizo and Velut 2006, 2020). However, energy market actors continue to work on the hypothesis of reinforced interconnections to accompany the energy transition (Velut 2021). Improved border crossings have facilitated the development of remote regions, which have found port outlets more easily and have also benefited the populations.

These circulate all the more easily because of the cultural and linguistic proximity, the weakness of controls in the border areas and the low level of regulation, such as the ability to move from one country to another without a visa. Today, the problem of migration to the United States is emphasized, made particularly visible by the caravans of migrants from Central America that cross Mexico before running into the border wall. Since 2018, the political and economic crisis in Venezuela has prompted migration on an unprecedented scale: an estimated two to three million Venezuelans have left their country, usually for other Latin American states. However, the importance of older migrations must also be stressed. There are well-established groups such as Bolivians in Buenos Aires, Peruvians in Santiago de Chile or Bolivians and Paraguayans in Brazil (Souchaud 2019). Before the current migration crisis, migration between Colombia and Venezuela went in one direction or another depending on the economic and political situation (Yepez Rousset and Garcia 2019).

Finally, a multitude of non-state actors, such as local governments, universities, trade unions and social movements, maintain continent-wide relationships. For example, the Mercociudades association created by municipalities in MERCOSUR has spread throughout Latin America. It promotes the exchange of best practices among local governments, with the aim of advancing integration through the intensification of relations between partner cities. Even though the resources devoted by each of these cities to decentralized cooperation are limited, Mercociudades brings to life the idea that Latin American cities share common challenges and must respond to them collectively without going through central governments. Similarly, the Montevideo Group, which brings together universities, initially limited, has expanded to Chile and Bolivia, with the aim of strengthening cooperation and mobility in the image of the European Erasmus program, but carried directly by the member universities. Several social movements are articulated in transnational networks, such as the movement of people affected by dams (MAB), which carry out concerted actions in different countries. This is also the case of indigenous movements, which organize transnational networks with the objective of bringing together communities belonging to the same ethnic groups, such as the Aymara, scattered between Bolivia, Peru, Bolivia and Argentina, and who recognize themselves in a common flag of the Andean peoples: the Wiphala. But it is also about bringing a common voice on the challenges of Latin America, especially around the environment and the rights of indigenous peoples, reactivating the idea of an America without borders (Paz Herrera 2016).

14.6. Conclusion: the meaning of integration

In Latin America, the notion of integration remains central to thinking about a large region, but its meaning is multiple. For states, integration can mean the creation of common political forums, the establishment of free trade areas or common markets, investment plans to improve connectivity and the creation of supranational political bodies. But these same states are anxious to preserve their independence. They refuse transfers of sovereignty and forms of coordination that are too restrictive, and often prefer alliances with major powers and commercial partners outside the region. For their part, civil society actors, local governments and businesses can design and implement sectoral or thematic integration projects, sometimes with the support of states, sometimes against their policies. The similarities between the countries, the social problems encountered and today's social and environmental struggle movements justify these connections.

Given the weight of political actors and representations, Latin American integration has received little attention from geographers from the region, with the exception of Brazilian geographers who regularly produce research on South America. It is mainly geographers from elsewhere who look at the continent and are most interested in the dynamics of integration. We can mention as examples the works of Perrier, Medina, Girault and Velut. Anglophone geography has taken another turn towards political ecology in a critical perspective.

But in general, the scientific production of geographers, wherever they are, is much less than that of economists, political scientists and international relations specialists, and they favor other scales of analysis. It was probably during the launch of the physical integration initiative that geographers became more involved (Costa 2009) or to address local situations of cross-border integration (Guibert et al. 2009; Rückert and Dietz 2013). Whatever the dynamics, they concern spaces smaller than Latin America: South America, Central America and the Caribbean form distinct sets. This is why the UNASUR initiative, which for the first time included the whole of South America, and the CELAC INITIATIVE, which proposes to bring together all the states south of the Rio Grande, are original, while there are still many sub-regional organizations with different aims.

Integration, which appeared with the States and in the same years as the expression Latin America, does not constitute in this great Region a post-state or post-national horizon, but an inseparable dimension of the constitution of States and Nations. The dynamic relationship between States, Nations and Integrations has not finished renewing itself and reconfiguring the territories of Latin America.

14.7. References

Amilhat-Szary, A.L. and Hamez, G. (eds) (2020). *Frontières*. Armand Colin, Paris.

Arab, C. (2009). *Les Aït Ayad : la circulation migratoire des Marocains entre la France, l'Espagne et l'Italie*. Presses Universitaires de Rennes.

Bahoken, F. (2021). Carte blanche sur la migration syrienne. *Carnets néocartographiques* [Online]. Available at: https://neocarto.hypotheses.org/12389.

Bataillon, C., Deler, J.-P., Théry, H. (1994). *Amérique latine*. Belin, Paris.

Carrizo, S. and Velut, S. (2006). L'énergie dans le Mercosur : réseaux transnationaux et logiques nationales. *Hérodote*, 123, 4.

Carrizo, S. and Velut, S. (2020). Régionalisations énergétiques sud-américaines. *Colloque du CIST*, Paris-Aubervilliers.

Cerrutti, M. (2020). 5 salient facts about intra-regional migration in South America. *Migration Data Portal*, 16 April [Online]. Available at: https://www.migrationdataportal.org/blog/5-salient-facts-about-intra-regional-migration-south-america.

Cortes, G. and Pesche, D. (2013). Territoire multisitué. *L'Espace géographique*, 42(4), 289–292.

da Costa, W.M. (2009). O Brasil e a América do Sul : Cenários geopolíticos e os desafios da integração. *Confins*, 7.

Dabène, O. (2014). La quatrième vague de régionalisme. *Les enjeux du développement en Amérique latine. Dynamiques économiques, insertion internationale et transformations sociales*, 64–95.

De Haas, H., Castles, S., Millerv M.J. (2014). *The Age of Migration*, 5th edition. Palgrave Macmillan, London.

Denis, P. (1927). *Amérique du Sud*. Armand Colin, Paris.

Diamint, R. (2013). Regionalismo y posicionamiento suramericano : UNASUR y Alba. *Revista CIDOB d'Afers Internacionals*, 101, 55–79.

Dorai, K., Hily, M., Ma Mung, E. (1998). Bilan des travaux sur la circulation migratoire. Summary report, Direction de la population et des migrations, Paris.

Droulers, M. (2001). *Brésil, une géohistoire*. Presses Universitaires de France, Paris.

Dumont, G.F. (2009). Le système migratoire méditerranéen. *Outre-Terre*, 23(3), 257–272.

Faret, L. (2003). *Les territoires de la mobilité : migration et communautés transnationales entre le Mexique et les États-Unis*. CNRS Editions, Paris.

Flahaux, M.L. and De Haas, H. (2016). African migration: Trends, patterns, drivers. *Comparative Migration Studies*, 4(1). doi: 10.1186/s40878-015-0015-6.

Grenier, P. (1988). L'effet-frontière dans l'utilisation et l'organisation de l'espace des Andes argentines. *Revue de géographie alpine*, 76(1), 7–44.

Guibert, M., Carrizo, S., Ligrone, P., Mallard, B. (eds) (2009). *Le Bassin du Río de la Plata : développement local et intégration régionale*. Presses universitaires du Mirail, Toulouse.

Héran, F. (2018). Comment se fabrique un oracle. La prophétie de la ruée africaine sur l'Europe. *La vie des idées*, 18 September [Online]. Available at: https://laviedesidees.fr/migrations-afrique-prejuge-stephen_smith-oracle.html.

Lambert, N. (2015). Une carte à abattre. *Carnets néocartographiques* [Online]. Available at: https://neocarto.hypotheses.org/1963.

Landau, L.B. (2019). A chronotope of containment development: Europe migrant's crisis and Africa reterritorialization. *Antipode*, 51(1), 169–186.

Mariano, K.P., Bressan, R.N., Luciano, B.T. (2021). Liquid regionalism: A typology for regionalism in the Americas. *Revista Brasileira de Política Internacional*, 64, 2.

Massey, D.S., Arango, J., Hugo, G., Kouaouci, A., Pellegrino, A., Taylor, J.E. (1993). Theories of international migration: A review and appraisal. *Population and Development Review*, 19(3), 431–465.

McAuliffe, M. and Khadria, B. (2020). *World Migration Report 2020*. IOM Publishing, Geneva.

Medina, L. (2019). Les projets de corridors interocéaniques en Amérique centrale : échelles et enjeux. *EchoGéo*, 49.

Migreurop (2017). *Atlas critique des migrations en Europe*, 2nd edition. Armand Colin, Paris.

Mützelburg, I. (2019). Régime migratoire. *Hypothèses mobilités, migrations, recompositions des espaces* [Online]. Available at: https://rm2.hypotheses.org/997.

Paz Herrera, M. (2016). Redes transnacionales de organizationes indigenas. Analisis del uso de las redes en conflictos socioambientales. *Revista de estudios sociales*, 55, 63–72.

Péraldi, M. (2001). *Cabas et containers. Activités marchandes informelles et réseaux migrants transfrontaliers*. Maisonneuve et Larose, Paris.

Perrier Bruslé, L. (2013). La Bolivie, sa mer perdue et la construction nationale. *Annales de géographie*, 689(1), 47–72.

Pfirsch, T. (2021). Les tours d'Europe familiaux des grands-parents italiens. In *Dossier "Quo vadis Europa ? La libre circulation européenne à l'épreuve des crises"*, Schmoll, C. (ed.). De facto [Online]. Available at: https://www.icmigrations.cnrs.fr/2021/04/27/defacto-026-05/.

Pison, G. (2019). Le nombre et la part des immigrés dans la population : comparaisons internationales. *Populations et Sociétés*, 563 [Online]. Available at: https://www.ined.fr/fr/publications/editions/population-et-societes/nombre-part-immigres-dans-population-comparaisons-internationales-2019/.

Recchi, E. (2021). Face aux crises, où en est la libre circulation européenne ? In *Dossier "Quo vadis Europa ? La libre circulation européenne à l'épreuve des crises*", Schmoll, C. (ed.). De facto [Online]. Available at: https://www.icmigrations.cnrs.fr/2021/04/27/defacto-026–02/.

Revel-Mouroz, J. (1972). *Mexique. Aménagement et colonisation du tropique humide.* Institut des Hautes Études de l'Amérique Latine, Paris.

Rückert, A.A. and Dietz, C.I. (2013). Integração regional, a região transfronteiriça da bacia do Rio da Prata e os projetos de infraestruturas de conexão. *Confins*, 17.

Schmoll, C. (2004). Une place marchande cosmopolite. Dynamiques migratoires et circulations commerciales à Naples. Geography PhD Thesis, Université Paris X Nanterre.

Schmoll, C. (2020). *Les damnées de la mer. Femmes et frontières en Méditerranée.* La Découverte, Paris

Scholvin, S. and Malamud, A. (2020). Is Brazil a geoeconomic node? Geography, public policy, and the failure of economic integration in South America. *Brazilian Political Science Review*, 14(2). doi: 10.1590/1981-3821202000020004.

Simon, G. (2002). Penser globalement les migrations. *Revue projet*, 272(4), 37–45.

Smith, S. (2018). *La ruée vers l'Europe. La jeune Afrique en route pour le Vieux Continent.* Grasset, Paris.

Sorre, M. (1928). *Mexique, Amérique centrale.* Armand Colin, Paris.

Souchaud, S. (2009). *Géographie de l'atelier : confection, migration, urbanisation à São Paulo.* Éditions de l'IHEAL, Paris.

Tarrius, A. (1992). *Les fourmis d'Europe. Migrants riches, migrants pauvres et nouvelles villes internationales.* L'Harmattan, Paris.

Théry, H. (2012). *Le Brésil*, 4th edition, Armand Colin, Paris.

Trenz, H.J. and Triandafyllidou, A. (2017). Complex and dynamic integration processes in Europe: Intra EU mobility and international migration in times of recession. *Journal of Ethnic and Migration Studies*, 43(4), 546–559.

UN DESA (2020). International migration highlights. United Nations [Online]. Available at: https://www.un.org/en/desa/international-migration-2020-highlights.

Valenta, M. (2020). Itinerant labour: Conceptualising circular, serial and stepwise migrations to the Arab Gulf and onwards. *Migration and Development.* doi: 10.1080/21632324.2020.1810897.

Van Houtum, H. and Bueno Lacy, R. (2020). The migration map trap. On the invasion arrows in the cartography of migration. *Mobilities*, 15(2), 196–219. doi: 10.1080/17450101.2019.1676031.

Velut, S. (2009). Argentine – Chili : une si longue frontière. *Confins*, 7. doi: 10.4000/confins.6095.

Velut, S. (2021). Une approche géographique des transitions énergétiques en Amérique latine. *L'Information géographique*, 85(3), 29–49.

Veyne, P. (1983). *Les Grecs ont-ils cru à leurs mythes ? Essai sur l'imagination constituante.* Éditions du Seuil, Paris.

Yepez Rousset, B. and Garcia, J. (2019). Le pull and push de la migration vénézuélienne. In *Venezuela : la revolution bolivarienne, 20 ans après*, Garzon, O.S., Sallerin, M.S, Carreno, E.U. (eds). Éditions L'Harmattan, Paris.

15

The People's Republic of China: Regional Pre-eminence as a Mirror of its Global Power

Thierry SANJUAN[1] **and Karine HENRIOT**[2]
[1] *UMR Prodig, Université Paris 1 Panthéon-Sorbonne, France*
[2] *UTC-Avenues, Sorbonne Université, Paris, France*

15.1. Introduction

China's economic and geopolitical emergence has made it the main player in the Asian region and, at the same time, the rising power on a global scale in relation to the United States. Looking at the Chinese case requires decentering the regional approach traditionally understood in the sense of a "geographical" region or in the sense of a regional cooperation agreement. This contribution illustrates a particular form of integration: the integration of a country within a region in which it sees itself as the major and central player. From then on, Chinese integration corresponds to a movement of self-integration and regional domination. It is this capacity to articulate regional development dynamics through a centripetal movement, and thus to polarize the Asian region, that allows China to position itself as a world power.

Since the beginning of the "reforms and opening up" (*gaige yu kaifang*) in 1978, this country–continent has undergone intense, rapid and even brutal upheavals in spatial, economic and social terms, which are directly linked to its integration into the global economic system. In addition to its trajectories of industrialization – initially coastal in the 1980s – then of urban catch-up and metropolization from the 1990s onwards, and now of sectoral tertiarization and innovation, the opening up of

Regionalization of the World,
coordinated by Pierre BECKOUCHE and Yann RICHARD.

China has benefited from its geographically central position, at the crossroads of Northeast Asia, the China Seas, Southeast Asia and Central Asia, as well as its links with economic actors of Chinese origin at the commercial helm of many Southeastern Asian countries.

In this, Beijing has relied on its regional integration to assert itself as a major player in Asia and thus potentially in the world, benefiting also from the joint economic dynamism of its neighbors (little dragons, baby tigers, etc.), with Hong Kong for a long time as the main hub of its opening. A common vision would have China reactivate its historical imperial primacy, vassalizing former tributary countries and erasing at the same time the Chinese collapse of the modern period. This view of impossible imperial mourning must be tempered by the modernity of the Chinese nation-state, and the training of its senior officials and diplomats to international standards.

It would be more appropriate to emphasize that Beijing is less inclined to integrate itself into its regional environment than to try to reconfigure Asian alliance networks. In this area, China has partners in mainland South and Southeast Asia (Myanmar, Pakistan), but it is also encountering resistance in Northeast Asia and in the China Seas. In the context of the Covid-19 crisis, its aggressive diplomacy, its maritime maneuvers, the issues of Hong Kong and Xinjiang, and its threats to Taiwan have recently changed the perception of China by other Asian countries. Above all, Beijing is confronting, both regionally and globally, the American rival and its allies.

In order to present these economic and geopolitical developments, our chapter is divided into three parts: the first part analyzes the end of isolation, regional integration and diplomatic strategies of mainland China with regard to its economic development; the second part focuses on the territories, tools and challenges of this Asian integration; finally, the third part considers China's global projection from its regional preeminence in Asia.

15.2. From regional integration to the construction of its own regional mechanism

The contemporary evolution of mainland China's ties with its regional environment involves three main stages. Firstly, it must break its hostile isolation from its Soviet neighbor and its allies on its land borders, as well as from the United States and the liberal Asian countries.

While the choice of "reforms and opening up" when Deng Xiaoping returned to power in 1978 had as its main objective a tacit contract with the Chinese population – for them the material improvement of living conditions, for the Party the maintenance of power – it also required the establishment of diplomatic links and the construction of conditions for economic integration (investments, technology transfers, relocations, exports, etc.) with Japan, the small dragons (Hong Kong, Taiwan, Singapore, South Korea) and its Southeast Asian neighbors, including the baby tigers (Thailand, Malaysia, Indonesia, the Philippines), as well as with Western markets. The average annual double-digit growth of its gross domestic product during the "Thirty Glorious Years" (1978–2008) allowed Beijing to integrate regionally and to establish itself as the main exporting power in the region.

After the Asian crisis of 1997 and the launch of the "going out" (*zouchuqu*) policy two years later, in which the central government invites large companies, small entrepreneurs and migrants to gain access to foreign markets around the world, the question posed to Beijing is no longer its integration into the Asian region, but rather the reconfiguration of this region to its advantage through alliance networks of alliances, multilateral institutions and the construction of partnership projects, under the label of the "New Silk Roads" – and more broadly that of "Asia for Asians" – from 2013 onwards, in order to counter the United States, and even the West as a whole, regionally and globally.

15.2.1. *Breaking out of isolation*

The isolation of the People's Republic of China (PRC), which was not recognized by the United States and international bodies in 1949, was aggravated by the Sino-Soviet split in 1960. Mainland China certainly wanted to be one of the mentors of the Third World, offering a path of development in Maoism, but it felt threatened regionally by its neighbors: conflicts with India in 1962 (Aksai Chin, Mac Mahon Line), the USSR in 1969 (Ussuri) and Vietnam in 1979; in the 1960s, China implemented the strategy of the "third line of defense" protecting the interior country from possible attacks by the USSR on land and the United States by sea.

Before the reforms themselves, the recognition of the PRC at the UN and its Security Council in 1971, at the expense of the Republic of China in Taiwan, and national policies that foreshadowed the Dengue period structurally changed the regional situation. Nixon visited Beijing in 1972, and the United States established formal diplomatic relations with Beijing on January 1, 1979. Japan, South Korea and other Asian countries normalized their relations with the PRC. At the same time, Zhou Enlai launched the "four modernizations" policy (agriculture, industry, science and technology, national defense), drafted by Deng Xiaoping, in 1975. Sino-foreign partnerships were already being set up, such as the one with the Federal Republic of Germany for the Baoshan steel complex north of Shanghai in 1977.

15.2.2. *Economic integration*

The Denguist reforms of 1978, which promoted the economic integration of China with its region, were based on three principles: deconcentration of decision-making powers in economic matters, thus leaving the initiative in development to the local levels (provinces, municipalities, districts, towns and townships); promotion of light industry, producing consumer goods for export; and preference for the coastal provinces and municipalities, particularly in the south (from Guangdong to Jiangsu), which are places of foreign investment, industrial relocation and export by sea.

Although China exported extensively to the markets of the developed countries of Asia (Japan, South Korea) and the West in the early days, the "opening up" was above all reflected in synergies with the extra-continental territories of Chinese settlement. Hong Kong played a central role, as it was both a source of relocation of production units, from the British colony of the time or via the Territory, essentially from the Southeast Asian countries where there was a Chinese diaspora eager to take advantage of the comparative advantages of the Continent, and a modern export port for the products of the enterprises of the boroughs and townships (*xiangzhenqiye*) of the Pearl River Delta in Guangdong (Sanjuan 1997). *Through* Hong Kong, Taiwan is also establishing industrial complementarity with southern Fujian.

In 2019, these dynamics – and hierarchies – at the heart of China's regional integration remained strong: 85% of the foreign direct investment (FDI) earning the country – out of a total of 138 billion dollars – came from Asia, of which 70% came from or passed through Hong Kong, 5% from Singapore, 1% from Macao, 1% from Taiwan – and concurrently 4% from South Korea and 3% from Japan. In the same year, China invested an equivalent amount of FDI worldwide (137 billion dollars), 81% of which was in Asia: 66% in Hong Kong, 4% in Singapore, 2% in Indonesia, 1% in Vietnam and 1% in Thailand (National Bureau of Statistics of China 2020).

15.2.3. *Building a regional environment of your own*

For a long time, Chinese diplomacy refused multilateralism, preferring bilateral negotiations where, thanks to the weight of the country, it could have the advantage. The Asian crisis of 1997 made China the arbiter of East Asia's economic equilibrium, and Beijing gave up devaluing the Yuan in order to limit the competition of its products with those of Southeast Asian countries. Under the guise of a policy of "good neighborliness", its new regional preeminence now allows it to enter into multilateral partnerships that institutionally extend its regional influence, and it now finds itself without a regional rival of its size: Japan's GDP was nine times greater than that of the PRC in 1990 but nearly three times smaller in 2018, according to the World Bank. Above all, China stands at the crossroads of major Asian subsets (Cabestan 2015, 2019).

The Shanghai Group, comprising China, Russia, Kazakhstan, Kyrgyzstan and Tajikistan, was formed in 1996 as a police and military organization. Five years later it became the Shanghai Cooperation Organization (SCO), with Uzbekistan as a member, India and Pakistan as observers, and Mongolia, Iran, Afghanistan and Belarus as observers.

In Southeast Asia, China is present through its investments and industrial relocations (Cambodia, Vietnam), its infrastructure projects (roads, railroads, bridges, canals, dams, etc.) that cross the Indochinese and Malayan peninsulas, initially within the framework of the Greater Mekong Region project – with the Asian Development Bank and Japan from 1992 – reclassified with the label of the "Belt and Road Initiative" (*yidai yilu zhanluë*) since 2013.

The Singapore summit in 1997 drew up the first elements of the ASEAN+3 perimeter (China, South Korea, Japan) with a view to a common free trade area. The Asian Infrastructure Investment Bank was created in 2015. Beijing is finally making its mark in the new Regional Cooperation Economic Partnership, which includes the ASEAN countries, China, South Korea, Japan, Australia and New Zealand, and represents 30% of world trade, in November 2020.

15.3. The territories of integration in Asia

The geography of Chinese integration into the Asian region is a testament to Denguistic pragmatism. The opening up was spatially and sectorally selective until the early 1990s. China then imported development tools such as the free trade zone model, which had made Taiwan successful. The southern coastal provinces (Guangdong, Fujian, Zhejiang and Jiangsu) also based their development on dynamic rural industrialization with an export focus.

Secondly, Chinese metropolization and the reactivation of urban networks are in line with known processes on the Asian and global scales, with the parallel settlement of border disputes that have remained unresolved for a century and that were inherited from the intrusion of European colonial empires into Asia (Russia in the northeast, the Russian and British Empires in the west, French Indochina in the south).

Since the 2000s, the construction of physical infrastructure (roads, railroads, etc.) and intangible infrastructure (5G), Chinese investments and relocations have connected China to its neighbors and projected its emerging power beyond its borders.

Finally, inter-state cooperation in maritime security and more general security is complicated by Beijing's increasing aggressiveness towards its neighbors.

15.3.1. *The territories of the Chinese opening*

The opening of China to Asia and the world runs from the beginning of the reforms in 1978 to their revival in 1992.

Four special economic zones were created in 1980, on the border with or in connection with Hong Kong (Shenzhen and Shantou), to the north and west of Macao (Zhuhai) and opposite Taiwan (Xiamen). In 1984, economic and technical development zones concentrated on 14 coastal cities, including Guangzhou, Shanghai, Beijing, Tianjin and Dalian. The following year, three coastal regions (including cities and countryside) were recognized as exemplary areas of regional integration in China: the Pearl River Delta, southern Fujian and the Yangtze River Delta. The entire coastline was opened up and Hainan Island became the 5th special economic zone in 1988.

After the relative isolation linked to the events of Tiananmen in 1989, the Dengue revival three years later completed regional integration: Beijing made official the opening of border cities and capitals of provinces or peripheral autonomous regions to Russia, Kazakhstan, Myanmar and Vietnam, and at the same time opened the provinces and municipalities of the Yangtze River basin to foreign investment, as far as Chongqing and Chengdu.

Finally, the inclusion of continental China in Asia is reflected in the development of large cities, especially coastal cities, illustrated in particular by the metropolization of the municipalities of Shanghai and Beijing, and by their increasingly intense links with other Asian metropolises, reactivating old networks between Asian coastal regions (Gipouloux 2009).

15.3.2. *Border regulations*

This economic integration is accompanied by Chinese recognition of most of the borders from the "unequal treaties" of the "century of shame" (1839–1949) (Colin 2011).

Only the borders with North Korea, Mongolia, Pakistan, Nepal, Myanmar and Laos had been agreed upon at the beginning of the communist regime. Those with the USSR, India and Vietnam were still pending at the end of the 1980s.

The neighboring Central Asian states and Beijing agreed on their borders in 1991. In northeastern China, Sino-Russian negotiations, conducted until 2004, resulted in the Amur and Ussuri rivers becoming border rivers. The memory of the 1962 conflicts with India still hampers Sino-Indian relations and was reactivated with the military events of summer 2020. Finally, while Hanoi and Beijing agreed on their land borders in the early 2000s, disagreement over maritime boundaries remains sharp and is part of the PRC's broader claims in the South China Sea.

15.3.3. *Chinese territories of integration outside the Mainland*

Since the 2000s, Chinese tourism, investment and offshoring have been connecting China to its Asian neighbors, especially towards ASEAN countries. Chinese and Hong Kong real estate FDI to ASEAN countries accounted for about one-third of global investment in the region from 2014 to 2017, and this share increased to 46% in 2019. This Chinese FDI far exceeded (between 7 and 10 times) that issued by Japan, the United States and the European Union.

These investments take the form of real estate and urban construction (Chinatown, condominiums, new cities, special economic zones), physical infrastructure construction (roads, railroads, stations, ports, airports, hydroelectric dams and dedicated engineering structures) and, since the second half of the 2010s, intangible infrastructure construction (5G, data centers, contactless payment terminals in public transport). These investments are being driven by Chinese companies, or even private individuals, to export their expertise and secure their capital, and are directed towards projects that spatially agglomerate a cluster of initiatives, particularly in special economic zones. These "Chinese territories of integration outside the continent", the fruit of private investment, are paradoxically to the benefit of Chinese diplomacy, even though the state is often outside of these projects.

15.3.4. *Regional cooperation and security*

China's policy of integration with Asia is a continuation of China's domestic policy of cultural, economic and security integration of the Western territories in the 2000s. It is first and foremost a policy of economic development to which is added an increasingly aggressive security policy. Inter-state cooperation in the area of security reflects China's ability to promote the development and integration of its territory, and then, in a reticular continuity, to organize the development of Asia and regional integration to its benefit.

The Shanghai Cooperation Organization, whose vocation was initially police and military based, became in the early 2010s a tool for stabilizing the Central Asian region by strengthening the fight against movements denounced as fundamentalist and separatist in the Chinese peripheries of Tibet and Xinjiang. A Central Asian continental corridor is structured, for example, from the rail and road connections between Xinjiang, historical Dzungary, Iran and Turkey, thus promoting a "transnational integration dominated by China" (Carriou 2018).

In 2013, the One Belt, One Road (OBOR) initiative, now the Belt and Road Initiative (BRI), is an economic, security and diplomatic priority. The semantic shift reflects China's desire to turn this policy serving both domestic and Asian interests into a global policy. These "New Silk Roads" are part of a continuing *soft power* effort, aimed in particular at developing countries and regions weakened by economic and social crises in Asia, Africa and Latin America (Lasserre et al. 2020).

To structure its development assistance, China is also creating its own banking institutions. It co-founded the Brics Development Bank in 2014, renamed the New Development Bank (NDB), the Silk Road Fund in late 2014 and the Asian Infrastructure Investment Bank (AIIB) in 2015.

The "New Silk Roads" that radiate throughout Asia are thus simply a continuation of the development policy of the Chinese West in the 2000s, in the continuity of the affirmation of its power and the integration of the Asian region by combining economic influence, military capabilities and influence, at the risk of regional tensions.

15.4. Projection in Asia and regional tensions

Since the end of the 2000s, China has been building a region of its own in Asia, in line with its domestic policy. To impose its way, Beijing has launched a diversification of the instruments of its power, certainly economic and security, by integrating diplomacy and influence.

Since 2007, China has been formalizing its soft power strategy, characterized by multiple initiatives, such as the organization of Asian and international events, or the opening of hundreds of Confucius institutes. The "compatriots" of Hong Kong and Macao and the overseas Chinese are also relaying this projection in Asia.

The Chinese game, resolutely paternalistic towards allied states and more willingly aggressive towards its rivals, reveals strong regional tensions, exacerbated and renewed in 2020 and 2021 by the global Covid-19 pandemic.

From now on, the diplomacy of the "New Silk Roads" imposes the "Beijing consensus" and casually handles the ambivalence of a discourse differentiated between Asian power and world power.

15.4.1. *International events in Asia*

At the turn of the 2010s, China inserted itself on the Asian and global scene by organizing cultural events and sports meetings of regional and global scope. While China's two main continental metropolises are participating in international events (Beijing hosted the Olympic Games in August 2008 and Shanghai organized the 2010 World Expo between May and October), China's major cities are positioning themselves in the Asian area of influence. The southern metropolis of Guangzhou hosted the 2010 Asian Games and Hangzhou hosted the 2022 Games. Secondary cities are also mobilizing: Changchun hosted the 2007 Asian Winter Games; Haiyang, the 2012 Asian Beach Games; Nanjing, the 2013 Asian Youth Games; Nanjing, the 2014 Asian Youth Games; Sanya, the 2021 Asian Beach Games; Shantou, the 2021 Asian Youth Games. By organizing these cultural and sporting events, China is demonstrating that it is also capable of producing soft power in the service of an identity and visibility, for Asia itself and beyond.

15.4.2. *"Compatriots" and the Chinese overseas*

This influence is relayed in Asia by the "compatriots" of Hong Kong and Macao and the overseas Chinese. Historically, nine-tenths of the Chinese diaspora came from the three southern provinces (Guangdong, Fujian and Hainan) and emigrated during the second half of the 19th century. Since the contemporary opening up, new waves of emigration from the northeast, as well as the numerous Chinese students abroad, have contributed to the diversification and renewal of this diaspora. The nationals of the special administration regions of Hong Kong and Macao, as well as the overseas Chinese, are renewing the links that the Chinese communities have with the Continent, to form a regional and global network of small investors conveying the economic influence and cultural influence of China, especially in the integration territories of Southeast Asia, such as the Chinatowns of Singapore (3/4 Chinese), Kuala Lumpur, Bangkok or Sihanoukville. Paradoxically, it is the "compatriots" of Hong Kong and Taiwan that put the PRC under the most pressure and incite it to reveal an aggressive game, a source of tension in Asia.

15.4.3. *Growing regional tensions*

In 2014, the "Umbrella Revolution" in Hong Kong and the "Sunflower Revolution" in Taiwan illustrated the rising tensions in Asia. From then on, Beijing has been steadily strengthening its economic, political, security and societal grip on the Hong Kong Special Administrative Region, culminating in the Comprehensive Security Law, which effectively ratifies the end of the special "one country, two systems" regime in June 2020, in the midst of the Covid-19 pandemic, while Asian and Western powers are (pre)occupied with managing the health crisis on their own territory. The PRC is now mobilizing the slogan "one country, two systems" by turning to Taiwan, urging it to "reunify" while threatening to "crush" any attempt at independence.

In the East China Sea, Beijing defies Japan and Korea, not hesitating to create in 2013 an Adiz (Air Defense Identification Zone), superimposing itself on other Adiz on the disputed territories of the Senkaku Islands under Japanese control (Diaoyu Islands for China) and Socotra Rock, claimed by South Korea.

Faced with the "string of pearls" strategy, a term coined by the United States in 2004 and formally rejected by the Chinese to designate the protection of Chinese trade routes in order to secure its energy supplies, points of friction are multiplying in Asia and between China and the United States (Amelot 2010). In the South China Sea, landfills are being created on several islands in the Paracel Archipelago and artificial islands are being built in the Spratlys, at the cost of an ecological disaster. Chinese naval bases, or those financed by China in foreign ports, are thus established along its main maritime supply route from the Middle East. Along this "necklace" are the Yulin naval base (Hainan Island), the Sansha air base (Yongxing Island, one of the Paracel Islands), the Zhubi air base, Meiji and Yongshu (three of the Spratly Islands), Sihanoukville (Cambodia), Kyaukpyu and Sittwe (Myanmar), Chittagong (Bangladesh), Hambantota (Sri Lanka), Gwadar (Pakistan), Doraleh and Obock (Djibouti), as well as Port Sudan (Sudan).

Faced with this "great wall of sand" (a sarcastic expression given by the United States in 2014), an exclusively strategic counter-project was built: the Indo-Pacific alliance. Implemented first by India and Japan, the Quadrilateral Security Dialogue (QSD, then Quad), reactivated in 2017, relies on an architecture of military cooperation allying the United States, Japan, Australia and India, to contain what is considered a threat of Chinese expansion. Since March 2020, New Zealand, South Korea and Vietnam have joined the alliance against the backdrop of cooperation against the effects of Covid-19, disseminated from China. In September 2021, Aukus, the latest Indo-Pacific military and security alliance, brings together the United States, the United Kingdom and Australia. Seen as complementary to the Quad, Aukus aims to equip Australia with nuclear-powered submarines (at the expense of a previous French–Australian arms contract), to counter China's rise in power, in the fields of undersea security, cyberspace, quantum computing and artificial intelligence.

15.4.4. *The ambivalence of a differentiated discourse between Asian power and world power*

Indeed, China is playing on its power in Asia to better establish its global power, particularly in Africa and Latin America, in the face of the United States and the European Union. If domestic policy is particularly articulated to its foreign policy, there are like two locks of the same policy, two distinct narratives, one for Asians, the other for Westerners. This ambivalence reflects its status as an Asian power and a world power. In the spring of 2021, China did not hesitate to use this double capacity of enunciation, to support Myanmar's junta at the regional level, while taking a certain distance at the international level.

At the same time, the diplomacy of the "New Silk Roads" is based on the "Beijing Consensus" (a geopolitical formula born in the United States in 2004), an alternative development model proposed by China to developing countries, particularly in Africa. The West and China have very different positions on the needs and methods of development assistance. The Western position is summarized in the "Washington Consensus", with free trade, anti-corruption, transparency, civil rights, democracy and human rights as its main approaches. China's diplomatic approach places great emphasis on non-interference (independence in internal affairs), as well as on a development mode that focuses first on physical infrastructure (roads, railroads, ports, bridges) and intangible infrastructure (5G and data centers), then on economic infrastructure (industries, mining and oil extraction) and finally on civic infrastructure.

China is becoming more than ever an ambivalent power, a “potentially revisionist power” (Lasserre et al. 2020), within its national territory, regionally and beyond. The Chinese Internet has rapidly become an Intranet, and then a platform for surveillance and control of the Chinese population (in this the collaboration of the Batx – Baidu, Alibaba, Tencent and Xiaomi – with the regime is decisive). Since the Covid-19 pandemic, health and security policies have promoted, in China as elsewhere in the world, the tracing of infected and uninfected populations by a QR-Code, the generalization of a vaccine passport and biometric facial recognition of masked populations. Thus, the deployment of 5G by Huawei, in Asia and beyond, favors the projection capacity of the Chinese Internet, at the risk of capturing, using and hacking the data exchanged. These infrastructures are becoming a vector of power and cyber offensive, more precisely of cyber-espionage and clandestine actions, prolonging China’s worrying rise to power.

15.5. Conclusion: the power of proximity

In 40 years, the People’s Republic of China has come out of isolation, to favor an economic and security integration allowing it to build an Asian region on its own.

The geography of Chinese integration in the Asian region is marked by a spatially and sectorally selective opening until the early 1990s. Then, it was accompanied by Chinese recognition of its land borders, while the disagreement over maritime borders was part of the PRC’s broader claims to the South China Sea. Since the 2000s, regional economic integration has changed scale to link “Chinese territories outside the continent” to mainland China, such as the special economic zones reproduced in Southeast Asia, combining real estate construction, tangible and intangible infrastructure, and mobilizing Chinese investment and urban expertise. In the continuity of its domestic policy, Beijing is reconfiguring the networks of regional security cooperation and alliances, in order to tie down neighboring states, allies or partners, to the construction of a region serving its interests. It is this ability to articulate and polarize the Asian region that allows China to position itself as a global power. The years 2020–2021, the Chinese management of the global pandemic, the question of Hong Kong and now Taiwan, mark a pivot, a paradigm shift. China is now a source of regional tension and is perceived as an aggressive Asian and global power.

This vision of China building a regional space on an Asian scale at its heel corresponds to a vision shared in French geography, which insists on the historical continuities, the territories of this integration, by privileging a cultural and geopolitical approach. English-language geographic literature also uses a reading of regionalization and regional integration, speaking of a “China-centric economic order", while adopting a more geographically segmented economic and security approach, sometimes focusing on Central Asia, Southeast Asia, East Asia or the Asia-Pacific.

15.6. References

Amelot, L. (2010). La stratégie chinoise du "collier de perles". *Outre-Terre*, 25–26(2–3), 187–198.

Cabestan, J.-P. (2014). *Le système politique chinois, un nouvel équilibre autoritaire*. Presses de Sciences Po, Paris.

Cabestan, J.-P. (2015). *La politique internationale de la Chine, entre intégration et volonté de puissance*. Presses de Sciences Po, Paris.

Cariou, A. (2018). Les corridors centrasiatiques des nouvelles routes de la soie : un nouveau destin continental pour la Chine. *L'Espace géographique*, 47(1), 19–34.

Colin, S. (2011). *La Chine et ses frontières*. Armand Colin, Paris.

Gipouloux, F. (2009). *La Méditerranée asiatique, villes portuaires et réseaux marchands en Chine, au Japon et en Asie du Sud-Est, XVIe-XXIe siècles*. CNRS Éditions, Paris.

Lasserre, F., Mottet, É., Courmont, B. (eds) (2020). *Les nouvelles routes de la soie ; géopolitique d'un grand projet chinois*. Presses universitaires de Québéc.

National Bureau of Statistics of China (2020). *China Statistical Yearbook 2020*. China Statistics Press, Beijing [Online]. Available at: http://www.stats.gov.cn/sj/ndsj/2020/indexeh.htm.

Sanjuan, T. (1997). *À l'ombre de Hong Kong, le delta de la rivière des Perles*. L'Harmattan, Paris.

Wong, J. (2013). A China-centric economic order in East Asia. *Asia Pacific Business Review*, 19(2), 286–296.

16

Southeast Asia, a Region?

Nathalie FAU
CESSMA, Université Paris Cité, France

16.1. Introduction

Located at the end of the Asian continent, from which it is separated by mountain chains originating from the Himalayan foothills, Southeast Asia forms the tropical end of the Asian continent (Antheaume 1995; De Koninck 2005). This region, located at the crossroads between two continents, the Asian continent and Oceania, and between two oceans, the Indian and Pacific Oceans, has long been considered in the west only in relation to India and China (Bruneau 2006) or as a simple "corner of Asia" according to the expression coined by Reclus (1883), taken up by the sociologist Mus (1977). In addition, the region has been grouped, in its entirety or for certain parts only, with larger groups, such as the Far East or East Asia. It has also been divided into its insular and continental parts. In the first two Universal Geographies of Malte-Brun (first third of the 19th century) and Reclus (last quarter of the 19th century), insular Southeast Asia is separated from the continental and attached to Oceania. In Vidal de La Blache's *Géographie Universelle*, published in the first third of the 20th century, this division no longer exists, but the whole of Southeast Asia is included in a much larger whole, the Asia of the Monsoons, which extends to India. If Southeast Asia is thus "an area of variable geography" (Durand 2000), it is mainly because of the absence, throughout history, of a political system that would have unified this region, and because of the very great cultural and religious diversity of the populations. To identify and characterize this diversity, Lombard (1990) refers to "strata", "layers of more or less ancient cultures, unevenly piled on top of each other and whose outcrops, somewhat like on geological maps, constitute the current reality" (Tertrais 2002). It is precisely

Regionalization of the World,
coordinated by Pierre BECKOUCHE and Yann RICHARD.

this diversity, this crossroads situation and the proximity of major powers that make defining Southeast Asia as a region complex. To assert the existence of a Southeast Asian region raises at least three major questions. The first concerns the perception of this region by the international community and by Southeast Asian populations. The second concerns the modalities and sometimes the contradictions of an institutionalized regional integration, whose main driving force is the shifted insertion of Southeast Asian countries into globalization. The last concerns the risk of internal disintegration of ASEAN and the risk of its dilution in a much larger Asian region.

16.2. Regional integration as a tool for international recognition

16.2.1. *From an exogenous regional name to the construction of ASEAN*

The term Southeast Asia is recent and western, and does not refer to some ancient myth that would suggest its unity, as for Europe. Its first occurrences dates back to the 1920s, and referred in the works of Austrian archaeologists and German ethnologists to the so-called "bronze drum civilization", centered in part on present-day Southeast Asia and distinct from the great civilizations of China and India. Its dissemination outside of academic circles dates back to the Pacific War (1943) and the creation by the British of the South East Asia Command, whose mission was to counteract the Japanese occupation (Tertrais 2014). In the aftermath of independence, the denomination Southeast Asia, of external origin, was gradually absorbed, reappropriated and reinterpreted by Southeast Asian scholars, institutions and political elites. Initially exogenous, the concept has become endogenous and the bearer of political projects, of which the construction of ASEAN is both the symbol and the pivot of a new representation and perception of regional unity.

Created in 1967, ASEAN originally brought together five states (Indonesia, Malaysia, the Philippines, Singapore, Thailand). Its objective was then exclusively political. In the context of the Cold War, which was particularly "hot" in Asia (Tertrais 2002; Foucher 2002), its first objective was to form a common front in the face of the risk of the spread of communism and to display a willingness to be neutral with regard to the powers (the United States, the USSR and China) that were holding the region in a vice-like grip (Boisseau du Rocher 1998). Its second purpose was to stabilize and strengthen fragile states that had just emerged from a more or less violent phase of decolonization (with the exception of Thailand, a country that had never been colonized). The linear borders of Southeast Asian states, imposed by colonial empires, and the adoption of the western model of the nation-state also run counter to local conceptions of sovereignty and the spatial patterns of pre-colonial political structures. ASEAN allows these states to gain international recognition and internal legitimacy. It is thus more a means of promoting national interests than of regional construction. It is revealing that its founding principle, respect for national sovereignty, at the origin of the principle of non-interference, has never been questioned. Finally, ASEAN promotes a process of regional reconciliation by easing tensions related to the drawing of state borders and the persistence of centrifugal forces in the peripheral areas. At the turn of the 1990s, the end of the bipolar world gave rise to another regional construction narrative based more on economic cooperation and enlargement, with the integration of Brunei (1984), Vietnam (1995), Myanmar and Laos (1997) and Cambodia (1999). Apart from Timor-Leste, all the countries of Southeast Asia, as delineated by extra-regional actors, are members of an association that is working to build a region whose modalities and modes of operation are detached from external models to assert a unique example of an Asian regional system (Nicolas 2017).

16.2.2. *The difficult emergence of a Southeast Asian community*

How do you create a region when the signs of economic and social diversities and disparities outweigh those of unity? It is difficult to find historical, linguistic, cultural or religious congruence in Southeast Asia. Far from minimizing these differences, ASEAN members have made them an asset and the very foundation of the *ASEAN way*, which is so different from the western approach to regionalization: pragmatism, the search for compromise and the lowest common denominator prevail over any theoretical consideration; discussion and process are often more important than the final decision (Nicolas 2017). This approach explains the refusal to set up a supranational community institution with its own powers that could impose binding measures on member states. The construction of a common identity is also based on economic success: the stability and prosperity of each country constitute a project, which, if not entirely collective, is simultaneous and parallel. In the 1980s and 1990s, local elites interpreted the rapid economic growth of Asian countries, which was so little predicted by the west, as the direct result of the characteristics of these societies: a hierarchical and unequal system based on age and seniority rather than on competition and rivalry, interpersonal relations that favor loyalty and the weight of family structures in economic activities (which also favors nepotism, corruption and the exclusion of those who do not belong to the group). Asian values, unlike those of the west, do not emphasize the individual and their aspirations, but the group and its prosperity. Based on the example of Japan, the aim was to challenge the universality of human rights and democracy advocated by the west.

ASEAN is also multiplying projects that promote the creation of an "ASEAN community". The ASEAN 2020 vision project, launched in 1997, aims to build an ASEAN community conscious of its ties in history, aware of its cultural heritage and bound by a common regional identity. Similarly, the 2003 ASEAN Charter sets the goal of creating an economic and security community, as well as a socio-cultural community. Conceived by heads of government and foreign ministers, the promotion of a common identity and regional citizenship has struggled to spread beyond the circle of intellectual and political elites. All surveys show that ASEAN is either not well known among the population or perceived as an elitist project. To gain notoriety, the institution has multiplied public information campaigns, incentives to visit member countries (free tourist visas for nationals of member countries), regional sports events (the ASEAN Games) and university exchange opportunities. While a change in mentality is undeniable, it varies according to age group, with young people feeling more concerned by ASEAN, and according to the level and mode of economic development: the populations of economically developed and liberal-oriented countries, such as Singapore, are more involved than poor populations for whom the institution has little direct impact.

However, the growing sense of urgency in the face of China's growing influence and the exacerbation of great power rivalries could play a unifying role and encourage the construction of a regional identity as a means of affirming ASEAN's centrality.

16.2.3. *The centrality of ASEAN*

Ongoing geopolitical changes have a direct role in transforming ASEAN's role and the limits of its scope of action. Faced with pressure from the great powers, ASEAN is asserting its "centrality" by ratifying agreements that integrate its major Asian partners (Boisseau du Rocher and Nicolas 2016). This institutional centrality was first manifested in 1994 with the creation of the ASEAN Regional Forum (ARF) dedicated to security issues in the Asia-Pacific region (27 members), and was affirmed after the 1997 crisis. In this context of crisis, revealing the economic and financial interdependence of Northeast and Southeast Asia, regional partnerships are once again a priority in order to establish mechanisms for a common response to global economic hazards. ASEAN is thus at the heart of the creation of the dialogue forum, known as ASEAN + 3 (China, Japan, Korea). In November 2012, ASEAN launched the Regional Comprehensive Economic Partnership (RCEP) at Indonesia's initiative. The RCEP includes, in addition to the 10 ASEAN member countries, six countries, namely those of ASEAN + 6 (China, Japan, South Korea, India, Australia and New Zealand), and excludes the United States. With this project, ASEAN wishes to put an end to the pile of bilateral free trade agreements of the ASEAN+1 type ratified with each ASEAN+6 partner. The objective is to create a vast free trade zone (3.5 billion inhabitants, i.e. 49% of the world's population and 30% of the world's GDP) in order to facilitate trade between countries that are increasingly linked by the fragmentation of production processes and the consolidation of global value chains. For ASEAN leaders, the RCEP should help support the region's economic growth, open up new markets and, above all, assert the weight of the institution and its partners as the center of gravity in Asia. It was ratified on November 15, 2020 by all of the prospective partners, with the exception of the Indian government, which is more interested in the projects of the United States and concerned about the growing influence of China in the region.

This mega-trade agreement competes directly with another project, supported until 2017 by the United States, the Trans-Pacific Partnership – TPP. The TPP, renamed CPTPP (Comprehensive and Progressive Trans-Pacific Partnership) in January 2018, brings together 12 countries on both sides of the Pacific. However, it splits ASEAN in two as only Brunei, Malaysia, Singapore and Vietnam are part of it. Until the withdrawal of the United States in 2017, the TPP was more a reflection of the United States' desire to maintain influence in Asia than an outline of regional construction.

16.3. Reconciling regional integration and integration into globalization

16.3.1. *An externally driven economic integration dynamic*

Unlike Europe, which is a regionalism of peer matches, regional integration in Asia is based on the association in the same economic dynamic of countries with similar growth trajectories (export promotion and strong extraversion strategies), but staggered in time (Chaponnière and Lautier 2016, 2018). This sequential development process and differences in economic levels led to successive waves of investment and industrial relocation from Japan to the Newly Industrialized Countries (South Korea, Taiwan, Hong Kong and Singapore) in the 1970s–1980s, then to the founding countries of ASEAN and finally to China, Vietnam, Cambodia and Laos. The regional division of labor, the complementarity and segmentation of production systems and the emergence of transnational value chains, particularly in the textile, electronics and automobile sectors, are directly responsible for strong economic interdependence and the weight of intraregional trade (Taillard 2004; Mottet and Jetin 2016): its share has hovered around 50% for the past two decades in East Asia (adding Northeast Asia and Southeast Asia). This integration has also been encouraged by legislative measures aimed at enabling companies to exploit complementarities between countries and facilitate trade; the 1998 Brand to Brand

Complementation (BBC) grants companies that join the scheme a reduction in customs duties on spare parts or intermediate products imported from other ASEAN countries. Toyota's use of this mechanism has enabled the development of regional cooperation in the automotive sector.

Economic integration is thus driven by multinationals, intra-firm networks, intra-industry trade and foreign direct investment (FDI), the stock of which in Southeast Asia in 2016 was equal to that of China and half that of Latin America, for comparison. This openness to FDI is partly due to the absence of local entrepreneurship, as well as to internal political strategies. While the European Union, the United States and Japan have long been the main investors in the region (with, respectively, 29%, 19% and 13% of the total over the period 1995–2000), it should be noted that intra-ASEAN investments have increased significantly since then, even exceeding those of the United States and Japan over the period 2014–2017. While the first phase of regional integration was dominated by Japan, due to massive relocations by Japanese companies, the second phase, which began in the aftermath of the 1997 crisis, was driven both by intra-ASEAN relocations and by the growing role of China. Intra-ASEAN trade now accounts for 25% of total ASEAN trade, up from 18% in 1980, and is thus much higher than that of other regional groups of southern countries such as MERCOSUR or SAARC. The stabilization at around 25% is due to the fact that regional integration is not only taking place at the ASEAN level, but also at the East Asian level. ASEAN+3 is thus a de facto integrated region.

16.3.2. *Re-engineering Southeast Asia by improving regional connectivity*

This economic integration is supported by improved transnational transport networks under the ASEAN Connectivity Plan. This plan, the Master Plan on ASEAN Connectivity or MPAC, was adopted in 2010 and is based on three pillars: the improvement of the institutional framework in order to reduce tariff and non-tariff barriers and thus promote the creation of a single market, the implementation of legislative mechanisms promoting greater mobility of people within ASEAN and finally, the development of transnational transport infrastructure, whose objective is to promote connectivity within ASEAN (Fau 2016). Since 2015, this plan has been replaced by the Master Plan for Connectivity 2025 (MPAC 2025) with similar objectives: to smooth intraregional traffic and bring transport infrastructure up to the same standards (Tertrais 2014).

The ports of Southeast Asia, four of which (Singapore, Port Klang, Tanjung Pelapas and Laem Chabang) are among the top 20 ports in the world, structure a meridian maritime route along the Pacific coast of Asia. Described as an "axis of growth" by Tertrais (2002), this maritime route is the backbone of a region larger than Southeast Asia and is referred to as the "Asia Pacific" or "Asian Mediterranean" (Gipouloux 2009). It is organized around several maritime basins interconnected by strategic straits (Malacca, Taiwan) and polarized by multimodal logistics platforms. This section of the circumterrestrial maritime route, which has a general east–west orientation, was a simple relay between Europe and North America when it was formed in the 1970s, but over time it has acquired its own dynamic, fueled by the growth of intra-Asian trade. This intra-Asian route is itself in flux (Fau and Tréglodè 2018): while half of its flows are still composed of trade between Northeast Asian countries, the most dynamic segments are meridian in orientation between Northeast Asia and Southeast Asia and intra-ASEAN. It has also become more complex, with an increase in the complexity of port hierarchies: the reticular organization of the 20th century, centered in the north on Japanese ports, in the south on Singapore and in the center on Hong Kong, has become denser with the emergence in East Asia of Korean and Taiwanese ports, then Chinese ports and, in Southeast Asia, those of Malaysia and Thailand, as well as, to a lesser extent, of Vietnam and Indonesia.

In the area of land infrastructure, the two flagship projects are the ASEAN Highway Network (AHN) project and the Singapore–Kunming Rail Link (SKRL). In all ASEAN countries, improving road infrastructure is a national priority thought increasingly in coordination with the networks of neighboring countries (Fau 2016). The improvement of the road network includes the construction of transnational axes. These corridors must promote the creation of a multipolar and transnational space, linking existing and emerging urban regions. It is in the mainland that this new model of economic development has been pushed the furthest, notably in the Greater Mekong Region or GMA (Taillard 2009; Fau, Khonthapana, Taillard 2014): born just at the end of the 1980s, the GMA brings together the five countries of the Indochinese Peninsula and two provinces in southern China (Yunnan having been joined at the end of 2004 by Guangxi). The Asian Development Bank (ADB), which is piloting and financing this project, has proposed to base regional integration on a revival of trade that was interrupted first by colonization, then by decades of war. The aim is to erase the divide between communist and liberal economy countries by transforming the battlefields into market-integrated spaces, thanks to transnational, meridian and transverse transport axes that now mesh the region.

The impacts of these connectivity projects on regional integration processes are ambiguous. In the development strategies of Southeast Asian governments, national priorities and strategies for insertion into global markets often take precedence over regional integration. In the maritime sector, there is growing competition among

ASEAN ports to capture the long-distance container flows that transit the region (Fau 2016). The result is a proliferation of port investments, often redundant, which are proving economically unprofitable and undermining complementarities between ports. Similarly, the creation of a single market in the maritime sector is made impossible by the proliferation of non-tariff barriers and by the refusal of many countries (Indonesia, Vietnam) to integrate cabotage flows in order to protect their domestic market from competition.

16.3.3. *The privileged territories of regional integration*

In the context of this regionalization, new spatial forms have emerged that transcend national borders. The most widespread form is that of cross-border economic development zones known as "growth triangles". The term "triangle" was first used on December 21, 1989, by Goh Chok Tong, then Deputy Prime Minister of Singapore, to refer to the Singapore-Johor in Malaysia-Riau in Indonesia cross-border economic zone or SIJORI (Fau 2015). The objective of this cross-border cooperation is to promote the economic development of three territories with different assets in order to form a larger economic basin with greater potential. The functioning of this area is based on the complementarity of the territories and an international division of labor on a micro-regional scale. Following the economic success of the SIJORI triangle, the Asian Development Bank (ADB) theorized and institutionalized this development model and contributed to its diffusion: the Northern Strait of Malacca Triangle comprising four states in the northern part of the Malay Peninsula, two provinces in northern Sumatra and 14 provinces in southern Thailand and the Bimp-eaga, the so-called "Greater East Zone", including Brunei Darussalam, the states of West Kalimantan and East Kalimantan and North Sulawesi in Indonesia, Sabah and Sarawak in Malaysia, Mindanao and Palawan in the Philippines. The objective is no longer so much to exploit complementarities as to unite in order to exist at the international level, to finance joint infrastructure programs and manage resources in common. These new transnational regional spaces are thus shaping an insular Southeast Asia organized around seas and straits, composed of spatial units structured by proximity relations (Charras and Franck 2000). In the continental part, this development model has also spread. In 1999, Vietnam proposed to its Indochinese partners, with the support of Japan, the creation of the Cambodia–Laos–Vietnam development triangle. This triangle encompasses the entire southeastern plateau of the Indochinese peninsula and is oriented towards the exploitation of hydroelectric, agricultural and forestry resources (Fortunel 2007). At the end of the 1990s, the ADB replaced or made the "triangles" more complex by introducing corridors, which are also planning tools designed to promote the development of the margins. The border areas located at the crossing points of the corridors are experiencing a new boom: industrial and commercial free zones, development of urban duplication (Lainé 2011, Franck 2014), growth of border tourism with the emergence of casinos and *resorts* (Tan 2011).

Cross-border and sub-regional cooperation initiatives have also multiplied in maritime areas. Joint development areas (JDAs) are cooperation zones that allow coastal states to jointly explore and exploit marine resources (mainly fisheries resources and hydrocarbons) located either on both sides of the border or in an area of overlapping territorial claims (Fau 2017). They are a way to maintain good political relations by avoiding exploitation conflicts, while promoting the sustainable management of marine resources. As early as the 1970s, the governments of Malaysia, Thailand, Vietnam and Cambodia successfully established JDAs in the Gulf of Thailand, a 300,000 km^2 semi-enclosed sea where maritime boundary delimitations are particularly contentious. In addition, even though ASEAN has not been able to establish a common policy for the environmental protection of marine areas, sub-regional initiatives have increased. The Sulu-Sulawesi Large Marine Ecosystem (LME), launched in 2001, is a cooperation between Indonesia, Malaysia and the Philippines to protect marine areas: in the framework of a tri-national committee, Indonesia is in charge of the sub-committee on the production of endangered marine species, the Philippines are in charge of marine parks and their networking and Malaysia are in charge of sustainable fishing (Cabanban 2018).

Often envisaged as catalysts for regional integration, these cooperation zones only promote exchanges and functional integration processes. They have not given rise to cross-border governance or institutional innovation, nor to the dynamics of collective projects, still less to a sense of belonging to a cross-border community. The border retains its role as a filter and control: the possibilities for workers to migrate depend on their country of origin and their qualifications. Single border control is struggling to be implemented and some borders are regularly closed, with the construction of walls sometimes even being considered (Thailand/Malaysia; China/Burma). Finally, the multiplication of these zones risks weakening the unity of ASEAN and reflects the rivalries between Southeast Asian States to build their political and economic hegemony and defend their areas of influence: Laos and Cambodia are thus involved in two competing projects, one driven by Thailand, the other by Vietnam (Fau et al. 2014). However, these risks of division are also fanned by the increasingly strong involvement of powers outside ASEAN.

16.4. Between risk of dilution and disintegration

16.4.1. *China, a factor in the disintegration of Southeast Asia?*

China's growing engagement on the international stage is coupled with geographic and cultural proximity in Southeast Asia. For China, ASEAN is an area of strategic influence and a space of resources and commercial opportunities. The ratification of the ASEAN-China Free Trade Agreement in January 2010, followed by China's "good neighbor" policy from 2013, have strengthened ties between the two partners. Under the Belt and Road Initiative (BRI), of the six strategic land route corridors, two involve Southeast Asia: the Bangladesh–China–India–Myanmar corridor and the China–Indochina corridor linking China to Thailand via the Lao section. One of the flagship projects is the Chinese financing of the Kunming–Boten–Luang Prabang–Vientiane railway line, which is to be extended to Bangkok with Thai investment (Mottet et al. 2015). However, the route of this line is a complete break with the strategies presented by the Asian Development Bank in its connectivity plan for Southeast Asia, and could thus be at the origin of a new organization of space in the mainland. This Chinese strategy, described as "rail diplomacy", raises three concerns. The first is that of financing, with the fear that Laos will become excessively indebted to its Chinese partners and that this project will put too much pressure on the country's public finances. The second is that of economic dependence. Economic cooperation between China and the beneficiary countries often results in the obligation to give priority to Chinese companies to implement projects financed by China. The last is that of a split in Southeast Asia between a continental Southeast Asia that would be in China's orbit and a more independent maritime Southeast Asia.

Southeast Asian countries have an ambivalent relationship with China perceived as both a source of economic opportunity and a threat, as evidenced by the persistent conflicts in the South China Sea (Côté and Martel 2014; Tréglodé and Frécon 2018). However, not all have the same relationship with this rising power. China's presence and influence are hegemonic in Laos, Burma and Cambodia: China is the leading investor there while also being a strong ally, even a shield, in the face of western criticism denouncing the abuse of these authoritarian regimes. As for the Philippines and Vietnam, their relations with China are particularly unstable and fluctuate according to governmental priorities: if the nationalist priority, i.e. the denunciation of China's actions in the South China Sea, results in a rupture of relations, the economic priority favors a rapprochement. Thailand, Malaysia and Indonesia are all increasing their economic relations with China while working to maintain their decision-making autonomy. This ambivalence is reflected in what political scientists call the hedging strategy: these countries refuse to align themselves with China, and try to avoid provoking it by allying themselves exclusively with a rival power (Delfolie et al. 2016). The growth of economic relations with China is, at the same time, part of a diplomatic tradition of balance between the great powers, particularly the maintenance of privileged relations with the United States. Finally, it should be noted that the development strategy chosen by the Singaporean government is a true model for China: a combination of economic liberalism in which the state remains very present and the hegemony of a single authoritarian party perfectly compatible with the Chinese system.

This diversity of relations between Southeast Asian countries and China is evident in the impossibility for ASEAN to take a common position on China's claims in the South China Sea (Mottet et al. 2017). The "Declaration on the conduct of parties in the South China Sea", ratified in 2002 by China and ASEAN in order to avoid violent conflicts, has not yet been transformed into a "Code of Conduct" that would be more binding for China. On the geopolitical level, the Chinese presence is upsetting older partnerships. In Laos, Vietnam and Thailand are losing influence. In Cambodia, in order to show allegiance to China, the government regularly makes political decisions unfavorable to Japan, even though Cambodia is one of the main recipients of Japanese aid. A split could thus emerge within ASEAN between a mainland part that is more oriented towards China and an island part that is multiplying its cooperation outside the region.

16.4.2. *A proliferation of regionalization projects larger than Southeast Asia*

The Chinese influence and presence must be qualified by the presence of other major influential powers in the region: first and foremost Japan and the United States, even though its retreat was evident during the Trump presidency, as well as, more recently, Australia and India. Thus, even though Chinese economic actors invest in many projects (resources, infrastructure), they are generally in competition with other states and private actors.

Rivalries between China and Japan crystallize in the mainland and in the construction of infrastructures. While China is dominant and omnipresent in Laos and Cambodia, Japan retains its influence in Vietnam. As for Burma, its allegiance depends directly on the evolution of its domestic politics: more oriented towards Japan when the government is civilian, it rallies to China when it is placed under a military regime (Nicolas 2018). Japan benefits from the anteriority of its involvement in the region (Boulanger 2001). It was behind the creation of regional organizations such as the Ministerial Conference for the Economic Development of

Southeast Asia and the Asia-Pacific Council. Moreover, in 1996, it founded the Asian Development Bank (ADB) with the United States. As its main shareholder, it directs the aid policies of this regional bank towards countries with which it has good relations. In the 1970s, Japanese aid to Southeast Asia exceeded American aid and was directed almost exclusively to the founding countries of ASEAN. Since 2015; however, the ADB has faced competition from the Asian Infrastructure Investment Bank (AIIB) promoted by China, which holds the presidency and a third of the capital. In the Greater Mekong Region, this competition is reflected in China's preference for financing north–south corridors and Japan's preference for east–west corridors (Taillard 2009).

These rivalries are spatially reflected in ASEAN's inclusion in strategic partnerships and networks that go beyond the boundaries of Southeast Asia. Since independence, Southeast Asian countries have been more oriented towards the Pacific than the Indian Ocean. This is by no means a historical constant, but a reversal after World War II, linked to the economic and political role of the United States in the reconstruction of non-communist countries. India's retreat into its national territory, the construction of ASEAN and then the rise of regional economic integration between the countries of Southeast and Northeast Asia had further distanced the Indian Ocean from the concerns of the countries of the "Asia Pacific": this regionalization, which refers to an economic reality, took on institutional consistency with the creation in 1992, on the initiative of Australia, of the Asia-Pacific Economic Cooperation (APEC), which now has 21 members bordering the Pacific Rim. Since the early 2000s, this "Asia-Pacific" has been gradually supplanted by a larger regional construction with a more strategic and geopolitical aim, that of the "Indo-Pacific". Launched by India in 2007 (Saint-Mézard 2014), and then taken up by Japan, this new name emphasizes the major role of maritime spaces in Asian international and strategic relations, as well as the growing interconnections between the Indian and Pacific Oceans. The term was then endorsed with its formalization in 2013 by Australia in its defense white paper, and then by Donald Trump's 2017 speech at the APEC summit: he spoke of a "free and open Indo-Pacific", explicitly recalling the "freedom routes" launched by India and Japan in response to China's "silk roads". In May 2018, US strategists replaced their military command in the Pacific (US Pacific Command) with an Indo-Pacific Command, covering the space from the west coast of the United States to the west coast of India.

But the contours of this region remain unclear; each government defines and delimits the region according to its own strategic interests. In order to maintain ASEAN's centrality and underscore their opposition to the explicitly anti-Chinese connotation of the term "Indo-Pacific", Southeast Asian leaders adopted their own

"Indo-Pacific vision" (ASEAN Outlook on the Indo-Pacific) in June 2019: more inclusive and with a regional focus, this approach is driven by Indonesia, whose leaders have been promoting its role as a "global maritime hub" since 2014 (Alles 2019; Germain 2018). This involves exploiting ASEAN's geographical centrality in the Indo-Pacific space to retain the initiative and relying primarily on the ASEAN regional framework to foster political and strategic cooperation in the region by opting for a now classic posture of "connecting connectivity" by promoting collaboration with other cooperation platforms such as the ASEAN Regional Forum (ARF), the ASEAN Defense Ministers Meeting plus (ADMM+), the East Asia Summit (EAS) and the Indian-Ocean Rim Association (IORA). The institutional regional organization of Asia is certainly a power and influence issue between the United States and China, but ASEAN, presented by political scientists as a weak and inefficient region, remains the heart of this integration.

16.5. Conclusion: the power of proximity

Three ongoing developments are noteworthy in order to understand the future of ASEAN as a region. The first is the shift from de facto regional integration, driven by transnational firms, to more formalized integration through free trade agreements (AFTA, as well as RCEP, which came into force on January 1, 2022, and CPTPP). These agreements, combined with strong growth in domestic demand, will promote the regional integration of Asian trade, while greatly reducing dependence on western countries. A study published in 2021 by the Asian Development Bank shows that Asia's regional value chains (a group that includes Southeast Asia as well as Northeast Asia, South Asia, Central Asia and Oceania) are now more regionally integrated than those of Europe: the share of regional value chains in global value chains is more than 70% in Asia, compared with 60% for Europe. This means that, on average, 70% of the final value of goods is produced in Asia, compared to only 60% for goods manufactured in Europe. The integration of regional value chains is particularly strong in the sectors of energy and raw materials, agricultural products, consumer goods and information and communication technologies. Similarly, while in 2000 the share of Asian regional trade was much smaller than in Europe (53.5% vs. 71.5%), the situations are getting closer today with a decrease on the European side (68.4% in 2020) against a steady progression in Asia (58.9% in 2020). In this context, ASEAN could become a simple component of a very large Asian region whose limits are still uncertain: for example, India remains outside these regional agreements at a time when cooperation projects between Southeast Asian countries and India are multiplying (maritime cooperation, the *Mekong-Gange program* between India and Thailand or the *Kaladan Multimodal Transit Transport Project* between India and Burma).

The second development is the positioning of ASEAN vis-à-vis Sino-American rivalries. While Asia is reorganizing itself economically without the United States, the strategic and geopolitical recompositions are determined by the "return" of this power to the region. Since the presidency of Barack Obama and the implementation of the so-called "pivot" strategy towards Asia, the United States has reaffirmed its presence in the Asia-Pacific region and manifested it through an increased military presence (the US Navy regularly conducts naval operations called "freedom of navigation" in the vicinity of islets controlled by China). The tripartite Aukus (Australia, the United Kingdom and the United States) defense agreement, ratified in September 2021, is not regional at all, given the great distance between the protagonists, but it focuses on one region, the Indo-Pacific, and could lead to regional strategic changes. Implicitly anti-Chinese, this alliance makes Southeast Asian governments fear that they will find themselves at the heart of a conflict that is not theirs and be forced to position themselves at the risk of imploding the ASEAN.

The last is the retreat of democracy in favor of authoritarian regimes supported by armed forces that criminalize political opposition. While economic liberalism is still the order of the day, it no longer translates politically into the spread of the western model of democracy. With the exception of Malaysia and Indonesia, political authoritarianism is becoming the norm: exceptional military rule in Thailand, authoritarian populism in the Philippines, military coups in Burma and the strengthening of single-party rule in Vietnam, Laos and Cambodia. In the face of these authoritarian excesses, ASEAN has shown itself to be voluntarily powerless in the name of non-interference in the domestic politics of its member countries. In this context, could new political forms of regional integration emerge? They could, in particular, come from civil society, which is mobilizing to protest against the decisions of governments. Cabasset et al. (2021) refer to a "transnationalization of civil society demands", which is manifested by the support of local populations for Asian migrants or by the use in Thailand, Burma, Taiwan and Hong Kong of the same rallying gesture during demonstrations: the three-finger salute, inspired by the American film *Hunger Games*.

16.6. References

Alles, D. (2019). Le sens de l'Indo-Pacifique : de l'ambiguïté sémantique à l'opportunité stratégique. *Les Grandes Dossiers de Diplomatie*, 53.

Antheaume, B., Bonnemaison, J., Bruneau, M., Taillard, C. (eds) (1995). *Asie du Sud-Est, Océanie*. Belin-Reclus, Paris, Montpellier.

Boisseau du Rocher, S. (1998). *L'ASEAN et la construction régionale en Asie du Sud-Est*. L'Harmattan, Paris.

Boisseau du Rocher, S. and Nicolas, F. (2016). Entre Asie orientale et Asie-Pacifique : la centralité de l'ASEAN à l'épreuve de la puissance ? *Interventions économiques*, 55 [Online]. Available at : http://journals.openedition.org/interventionseconomiques/2817.

Boulanger, É. (2001). La nouvelle politique commerciale du Japon et le régionalisme asiatique. *Cahier de recherche*, 1, CEIM, Université du Québec à Montréal.

Bruneau, M. (2006). *L'Asie d'entre Inde et Chine. Logiques territoriales des États*. Belin, Paris.

Cabanban, A. (2018). Le grand écosystème marin Sulu-Sulawesi. In *Mers d'Asie du Sud-Est. Coopérations, intégrations et sécurités*, Fau, N. and de Tréglodé, B. (eds). CNRS Editions, Paris.

Cayrac Blanchard, F., Defert, G., Durand, F. (eds) (2000). *Indonésie contemporaine. Hommage à Jacques Leclerc*. L'Harmattan, Paris.

Chaponnière, J.-R. and Lautier, M. (2016). L'intégration économique régionale en Asie du Sud-Est : une dynamique impulsée de l'extérieur. *Mondes en développement*, 175(3), 113–130.

Chaponnière, J.-R. and Lautier, M. (2018). *Économies d'Asie du Sud-Est. Au carrefour de la mondialisation*. Bréal, Paris.

Charras, M. and Franck, M. (2000). Quarante ans d'introversion de l'Indonésie ; l'éclipse de toute une région. L'ASE insulaire dans les recompositions spatiales. In *Indonésie contemporaine. Hommage à Jacques Leclerc*, Cayrac Blanchard, F., Defert, G., Durand, F. (eds). L'Harmattan, Paris.

Côté, D. and Martel, S. (2014). La Chine et l'Asie du Sud-Est : une relation ambivalente. *Monde chinois*, 38–39(2), 48–65.

De Koninck, R. (2005). *L'Asie du Sud-Est*. Armand Colin, Paris.

Delfolie, D., Fau, N., Lafaye De Micheau, E. (2016). *La Malaisie et la Chine, une relation précieuse*. Carnets de l'Irasec, Bangkok.

Durand, F. (2000). L'Asie du Sud-Est, une aire à géographie variable. *Limes*, 5, 184–193.

Fau, N. (2015). Johor Bahru, arrière-pays ou pôle émergent de la métropole transfrontalière singapourienne. In *Territoire de l'urbain en Asie. Une nouvelle modernité ?*, Manuelle, F. and Thierry, S. (eds). CNRS Editions, Paris.

Fau, N. (2016). Investment in infrastructure and regional integration: Will connectivity reduce inequalities? In *ASEAN Economic Community: A Model for Asia-Wide Regional Integration?*, Jetin, B. and Miki, M. (eds). Palgrave Macmillan, Basingstoke.

Fau, N. (2017). Les zones communes de développement dans le golfe de Thaïlande : un modèle de coopération ? In *Géopolitique de la mer de Chine méridionale. Eaux troubles en Asie du Sud-Est*, Mottet, É. and Lasserre, F. (eds). Courmont Barthélémy, Presses de l'Université du Québec.

Fau, N. and de Tréglodé, B. (2018). *Mers d'Asie du Sud-Est. Coopérations, intégration et sécurité.* CNRS Editions, Paris.

Fau, N., Khonthapana, S., Taillard, C. (2014). *Transnational Dynamics and Territorial Redefinitions in Southeast Asia: The Greater Mekong Subregion and Malacca Strait Economic Corridors.* Iseas, Singapore.

Fortunel, F. (2007). Le plateau des Boloven et la culture du café : entre division interne et intégration régionale. *L'Espace géographique*, 3, 215–228.

Foucher, M. (ed.) (2002). *Asies nouvelles.* Belin, Paris.

Franck, M. (2014). Twin cities and urban Pairs, a new level in urban hierarchies structuring transnational corridors? A case study of the Pekanbaru-Dumai urban pair. In *Regional Integration in Southeast Asia*, Fau, N., Khontaphane, S., Taillard, C. (eds). ISEAS, Singapore.

Germain, V. (2018). L'ASEAN est-elle soluble dans "l'Indopacifique" ? *Asialyst* [Online]. Available at: https://asialyst.com/fr/2018/11/07/asean-soluble-indo-pacifique/.

Gipouloux, F. (2009). *La Méditerranée asiatique. Villes portuaires et réseaux marchands en Chine, au Japon et en Asie du Sud-Est : XVIe-XXIe siècle.* CNRS Editions, Paris.

Khonthapane, S. and Taillard, C. (eds) (2014). *Transnational Dynamics and Territorial Redefinitions in Southeast Asia: The Greater Mekong Subregion and Malacca Strait Economic Corridors.* ISEAS, Singapore.

Lainé, E. (2011). Les villes frontalières thaïlandaises dans la régionalisation. *Moussons*, 18, 51–75.

Lombard, D. (1990). *Le Carrefour javanais. Essai d'histoire globale.* Editions de l'EHESS, Paris.

Mikic, M. and Jetin, B. (eds) (2016). *ASEAN Economic Community: A Model for Asia-wide Regional Integration?* Palgrave Macmillan, Basingstoke.

Mottet, É. and Jetin, B. (2016). L'Asie du Sud-Est et les chaînes de valeur. In *Christophe Colomb découvre enfin l'Asie. Intégration économique, chaînes de valeur et recomposition transpacifique*, Boulanger, É. and Mathieu, A. (eds). Athéna éditions, Outremont, Quebec.

Mottet, É., Courmont, B., Lasserre, F. (eds) (2015). *La Chine et le Monde. Quelles nouvelles relations, quels nouveaux paradigmes ?* Presses de l'Université du Québec.

Mottet, É., Lasserre, F., Courmont, B. (eds) (2017). *Géopolitique de la mer de Chine méridionale. Eaux troubles en Asie du Sud-Est.* Presses de l'Université du Québec.

Nicolas, F. (2017). La communauté économique de l'ASEAN : un modèle d'intégration original. *Politique étrangère*, 2, 27–38.

Nicolas, F. (2018). Catching up or staying ahead: Japanese investment in the Mekong region and the China factor. *Asie Visions*, 99, IFRI.

Saint-Mézard, I. (2014). L'Inde et l'intégration régionale en Asie orientale : l'arrimage progressif d'un acteur périphérique. *Monde Chinois*, 37(1), 48–57.

Taillard, C. (ed.) (2004). *Intégrations régionales en Asie Orientales*. Les Indes Savantes, Paris.

Taillard, C. (2009). Un exemple de régionalisation transnationale en Asie Orientale : le corridor de la région du Grand Mékong. *L'Espace géographique*, 38(1), 1–16.

Tan, D. (2011). *Du communisme au néolibéralisme : le rôle des réseaux chinois dans la transformation de l'État au Laos*. CERI-Sciences Po, Paris.

Tertrais, H. (2002). *Asie du Sud-Est. Enjeu régional ou enjeu mondial ?* Gallimard, Paris.

Tertrais, H. (2014). *Atlas de l'Asie du Sud-Est. Les enjeux de la croissance*. Autrement, Paris.

Tréglodé, B. and de Frécon, É. (2018). La Chine et l'Asie du Sud-Est, vers un nouvel ordre régional ? *Monde Chinois*, 54–55(2), 6–11.

Conclusion

For a More Comparative Geography

Pierre BECKOUCHE[1] and Yann RICHARD[2]

[1] *UMR Ladyss, Université Paris 1 Panthéon-Sorbonne, France*

[2] *UMR Prodig, Université Paris 1 Panthéon-Sorbonne, France*

The regionalization of the world is a fundamental transformation of the organization of space, because relations between human groups are still partly determined by geographical proximity, beyond the level of states that are now overwhelmed by all kinds of mobility: commercial exchanges, international migrations, the development of major regional infrastructures, and the co-management of certain cross-border public goods in more cooperative forms of governance. Exchanges as fluid and immaterial as finance and cyberspace are no exception to this regionalization. Neither are geopolitical issues, since security and conflict regulation are often easier to implement on a regional scale than on a global scale: it may, in fact, be easier to share certain collective preferences between neighbors than on a vast global scale. From this point of view, regionalism can be a lever for international socialization and a credible alternative scenario to the new bipolar Sino-American world or to the global metropolitan archipelago.

Regionalization is taking place in all parts of the world. The European Union promotes deep integration through the sharing of norms, through social, economic and political convergence, and through the creation of common institutions; however, it is only one model among others. In the Americas, two logics exist side by side. On the northern side, NAFTA and its little brother, the ACEUM, are making a powerful contribution to the integration of North America into a vast regional system dominated by the United States, whose influence is felt in the South.

In the South, partly under the impetus of Brazil, Latin American countries are attempting regional experiments with uneven success, while remaining firmly attached to the principle of national sovereignty. In East Asia, too, nationalism is still a structuring factor, yet the functional regional integration of East and Southeast Asia is proceeding rapidly, since it is based on economic complementarities between nearby countries. Chinese activism is both accelerating and slowing down this regional integration (see Chapter 15 by Sanjuan and Henriot). The difficulties of regionalization in Africa are greater. There is a tangle of regional organizations whose track record is often poor, except in a few areas (security, business law, infrastructure construction, etc.).

The form of the macro-region has indeed become a universal figure for the organization of space, albeit with highly different realities from one case to another. Regionalization is even being trialed in parts of the world where physical constraints make it hardly conceivable, as in the Arctic (see Chapter 12 by Lasserre and Escudé). Finally, underneath or alongside all the government-led initiatives, there are more discreet forms of regionalization that escape institutional logic; carried out by other actors, often "from below", they are also levers of spatial integration at higher scales.

C.1. The macro-region remains more of a level than a territory

By interposing themselves at an intermediate level between the states and the world, macro-regions have imposed themselves as an additional level. This is the conclusion reached by analyses of the regionalization of flows, economic exchanges (see Chapter 5 by Van Hamme) and the logistical organization of world space. Similarly, migration routes have become globalized, but the vast majority remain regional and South–South (see Chapter 2 by Schmoll), contrary to what the preconceived idea of a massive migration flow from the South to the North may suggest. Similarly, and this is counter-intuitive, financial actors operating in a sector of activity that is globalized sometimes construct forms of regional grouping (see Chapter 6 by Sainteville).

However, in most cases, there is no cultural and social appropriation of these macro-regional spaces, nor is there any supranational power (see the case of Southeast Asia in Chapter 16 by Fau). It is true that the political and institutional agreements reflect an active regionalism, but often formal or liberal (superficial regional integration), and in any case far from a regional government. The European Union is an exception, with supranational institutions and common legal rules that eventually weave a dense normative space. But these European mechanisms are put at the service of a very open economic space with little political control. Let us recall, for example, that the EU is a common market and a commercial actor, but that it is commercially open to the rest of the world and that it has never implemented a common economic policy – let alone a common defense. Whether February 24, 2022, marks a turning point in this respect will take some time.

Although they have imposed themselves in the organization of world space, large regions rarely become *territories* in the strong sense that would have reached a degree of integration high enough to become a reference point, a reason for collective commitment[1], an object of representation shared by the population living there and by external actors[2]. The representations are essential, including the fabrication of a common narrative, if necessary through the control of information as shown by the *runet* (see Chapter 7 by Limonier). Media geography is a powerful tool for analyzing these regional representations in the making (see Chapter 10 by Grasland, Laporte and Toureille).

All in all, we have the impression that the region remains more a level of functional organization of space than an appropriate territory. However, several contributions in this book show that this work of appropriation, either by the populations or by the political or economic powers, is underway. We see this with the growing concern for regional security (see Chapter 8 by Chauvin) in an increasingly broad sense: security in the strict sense, but also security of food or energy supplies (see Chapter 3 by Palle). We see this with the global geopolitical tensions which, together with the questioning of liberal globalization, reinforce the interest in shorter circuits and the relevance of the regional approach taking advantage of the neighborhood, although in a contradictory way in the case of East Asia because of the fear that China inspires in its neighbors. This can be seen in the deepening of European integration, the most advanced example of regional integration in the world, but so complex that we wonder whether European citizens are not building a common territory (circulation, norms, currency, etc.) without realizing it, a sort of unthought territory that they do not appropriate (except when they become aware of it on the occasion of a major external shock such as the Russian invasion of Ukraine).

In any case, and even if to varying degrees, the change in status of the neighbor can be seen in (almost) all of the contributions in the book, whether as a result of

1. A "desire to live together" or, at the very least, an awareness of a "duty to live together", to use the words of the economist Henri Regnault (2008).

2. What the political scientists Hettne and Söderbaum call *regionness* (Hettne and Söderbaum 1998), that economic, institutional, political and cultural substance that is sufficient to be finally appropriated by the actors.

increased cross-border relations, shared issues or a sense of belonging to a larger regional whole.

C.2. Questions to be explored

This book sheds light on questions that have not been sufficiently addressed. Chapter 9 by Magrin and Ninot shows the limits of regionalization in Africa in the context of extra-African economic domination and institutional weakness of states. Regionalization is partly carried out against the state because of informal and even criminal cross-border trafficking. The general idea that the future lies with macro-regions is mitigated by African realities, which remind us that the constitution of solid states remains a prerequisite. Moreover, there can be both regionalization of flows and regionalism, and also weak regional integration when flows to the rest of the world far outweigh exchanges within the region. China's mega-infrastructural approach to Africa may well enhance the relevance of the African Continental Free Trade Area, but with the risk of increased extraversion. Macro-regionalization and development do not necessarily go hand in hand.

Several contributions raise the question of the relationship between regional integration and social cohesion. The example of East Asia shows that regional integration can fuel the disconnection between the elites (favorable to regionalism) and the rest of the populations (more willingly nationalistic). This would then constitute an essential limit to an appropriation of this macro-regional space to make it a territory in the full sense of the term. Chapter 2 by Schmoll and Chapter 13 by Girault show that human mobility is a revelation of the integration and/or fragmentation dynamics of regional and interregional (in the Euro-Mediterranean case) spaces. Human mobility, in all its components (asylum, climate refugees, professional mobility, so-called permanent migration due to long-term settlements), is a privileged observation point for the degree of cohesion of these immense regional territories. The question remains: can there be real regional integration without free movement of people?

C.3. Delays and contributions of francophone geography

Francophone geography has been interested in each of the levels of spatial organization: the local, the national, the world and, between these last two levels, the macro-regions. One of its strong points is the analysis of the great regions of the world, starting with Europe. It has also been interested in the interaction between these levels, particularly between metropolization and globalization, or between metropolization and regionalization: transnational urban networks organize the space of large regions through logistical relations and the development of international corridors. Or through multilevel governance, especially in the European Union (Baudelle and Mérenne-Schoumaker 2011; Palle and Richard 2021). While this should be one of its multiscalar fields of study, francophone geography is less fruitful in the study of interactions between regions as well as between regions and the world, whose approaches are left to economists and political scientists in their studies of interregionalism.

On the whole, the fact that francophone geographers are observing the regionalization of the world is a rather original approach in the humanities and social sciences: historians are less interested in this theme and it is rather marginal among sociologists, with the exception of political sociology. In the field of mobility, for example, francophone geographers have played an important role in the reflection on the regionalization of migrations, nuancing the grand narrative of globalization, which continues to underestimate the place of intraregional migrations.

Another strong point of francophone geography in recent decades is interdisciplinary cooperation. To take the example of mobility, it is their interdisciplinary approach that explains the interest of geographers in the social (migratory networks), historical (sometimes post-colonial) and political (European integration, borderization) dimensions of the regional configurations analyzed. Most of the contributions in this book are based on interdisciplinary work.

For the time being, nevertheless, regionalization is a field of research in which geographers are lagging far behind other disciplines such as political science and economics[3]. The books and academic articles of geography on this theme are rare, their distribution remains confidential and there are few research teams specialized in this field.

The relative erasure of geography in general, and francophone geography in particular, is noticeable even in parts of the world where regional integration is a political issue of primary importance, such as in Latin America. There, it is once again economists and political scientists who occupy the field of academic debate. Some geographers work on South America, but not on Latin America. And those who study regional integration mostly measure the impacts of a particular project in

3. For example, Sabine Saurugger's recently republished textbook on *Theories and Concepts of European Integration* focuses on political science and political sociology. Economics produces basic textbooks on regionalism and regional integration in general, including undergraduate textbooks.

their country or in a particular sub-region (Rückert and Dietz 2014). Paradoxically, it is geographers from elsewhere who look at Latin America and are more interested in the dynamics of integration, especially francophone geographers (see Chapter 14 by Velut), while anglophone geography has turned to political ecology and favors critical approaches.

Geography in general, and francophone geography in particular, is not more influential in the study of North American integration. The European question mobilizes a lot of francophone geography but in a rather fragmented way, which hinders its heuristic contribution. Finally, few works approach the Runet from a geographic perspective, even though other disciplines mobilize spatial terminology to address the sovereign Runet by speaking of "territorialities" or "digital borders".

Another weakness, and although it has been studied by geographers for a long time, the regionalization of the world remains a secondary subject in geographic research itself[4]. More annoyingly, francophone geographers interested in this theme have not produced an autonomous conceptualization of the dialectic between regionalization and globalization, their approach remaining in the wake of English-language debates. Nevertheless, francophone geography can draw on interesting methodological achievements for the analysis of the geographical divisions of the world.

C.4. Method for a more comparative regional geography

The most daunting problem, which runs throughout this book, is that of method. If geographers are to move beyond their still strong monographic tradition, an effort must be made to increase generality and conceptualization beforehand.

This presupposes agreement on comparable quantitative and qualitative methods. They are often comparable in the reality of the various studies, but the formalization of a true comparative framework remains to be done. For example, to measure the effects of major logistics corridors on regional integration – how, beyond the mapping of transport networks, can we measure their structuring territorial effects (see Chapter 4 by Beyer)? Another aspect of this methodological question asks: are geographers capable of developing a regional integration index?

This could nurture, in a better coordinated way, the immense tradition of francophone geography in the area of regional studies. This would help it take its

4. It must be said that regionalization is also relatively ignored in the *mainstream* of the major English-language geography journals, with a few exceptions.

full place vis-à-vis non-francophone geography as well as other scientific disciplines.

C.5. References

Baudelle, G. and Merenne-Schoumaker, B. (eds) (2011). *Le développement territorial en Europe : concepts, enjeux et débats*. PUR, Rennes.

Palle, A. and Richard, Y. (2021). Multilevel governance or scalar clashes: finding the right scale for EU energy policy. *Tijdschrift voor Economische en Sociale Geografie*, 113(1), 1–18.

Rückert, A. and Dietz, C. (2014). Intégration régionale, la région du bassin du Río de la Plata (Amérique du Sud) et les projets d'infrastructures de connexion. In *La régionalisation du monde. Construction territoriale et articulation global/local*, Gana, A. and Richard, Y. (eds). Karthala, Paris and IRMC, Tunis.

List of Authors

Pierre BECKOUCHE
UMR Ladyss
Université Paris 1 Panthéon-Sorbonne
France

Antoine BEYER
UMR MATRIS
CY Cergy Paris Université
France

Emmanuel CHAUVIN
UMR LISST
Université Toulouse Jean Jaurès
France

Camille ESCUDÉ
CERI-Sciences Po
Paris
France

Nathalie FAU
CESSMA
Université Paris Cité
France

Christian GIRAULT
CNRS CREDA "Amériques"
Condorcet Campus
Aubervilliers
France

Claude GRASLAND
UMR 8504 Géographie-cités
Université de Paris
France

Karine HENRIOT
UTC-Avenues
Sorbonne Université
Paris
France

Antoine LAPORTE
UMR 5600 EVS
École Normale Supérieure de Lyon
France

Frédéric LASSERRE
Université Laval
Quebec
Canada

Kevin LIMONIER
Institut Français de Géopolitique
Université Paris-8
Saint-Denis
France

Géraud MAGRIN
UMR Prodig
Université Paris 1 Panthéon-Sorbonne
France

Olivier NINOT
CNRS
UMR Prodig
Aubervilliers
France

Angélique PALLE
Institut de Recherche Stratégique
de l'École Militaire
Paris
France

Yann RICHARD
UMR Prodig
Université Paris 1 Panthéon-Sorbonne
France

Maude SAINTEVILLE
Lycée Louis-le-Grand
Paris
France

Thierry SANJUAN
UMR Prodig
Université Paris 1 Panthéon-Sorbonne
Paris
France

Camille SCHMOLL
EHESS/Géographie-cités
Institut Convergence Migrations
Paris
France

Etienne TOUREILLE
UMR 6266 IDEES
Université de Rouen – Normandie
France

Gilles VAN HAMME
Université Libre de Bruxelles
Belgium

Sébastien VELUT
EHESS/Géographie-cités
Institut Convergence Migrations
Paris
France

Index

G, I

L, M

N, P

R, S

T, U

V, W

Printed and bound by CPI Group (UK) Ltd, Croydon, CR0 4YY
13/09/2026
14961069-0005